EVERY BRAIN NEEDS MUSIC

EVERY BRAIN NEEDS MUSIC

The Neuroscience of Making and Listening to Music

LARRY S. SHERMAN AND
DENNIS PLIES

With Illustrations by Susi B. Davis

Columbia University Press
New York

Columbia University Press
Publishers Since 1893
New York   Chichester, West Sussex
cup.columbia.edu

Library of Congress Cataloging-in-Publication Data
Names: Sherman, Larry S., author. | Plies, Dennis, author.
Title: Every brain needs music : the neuroscience of making and listening to music / Larry S. Sherman and Dennis Plies.
Description: [1.] | New York : Columbia University Press, 2023. | Includes bibliographical references and index.
Identifiers: LCCN 2022037355 | ISBN 9780231209106 (hardback) | ISBN 9780231557795 (ebook)
Subjects: LCSH: Music—Physiological aspects. | Music—Psychological aspects. | Neurophysiology. | Brain—Localization of functions.
Classification: LCC ML3820 .S44 2023 | DDC 781.1—dc23/eng/20220809
LC record available at https://lccn.loc.gov/2022037355

Cover design: Mary Ann Smith
Cover illustration: Shutterstock

We dedicate this book to our partners in life,
Diane and Barbara,
to our children,
Benjamin, Claire, and Alanna,
Matt, Dan, Melanie, and Rachel,
and to
teachers and students everywhere.

CONTENTS

PRELUDE

A NEUROSCIENTIST AND A MUSICIAN WALK INTO A ~~BAR~~ GYM

BEFORE WE move to the overture, allow us to introduce who we are and how this book came to be. It started when a neuroscientist and a musician walked into a gym very early one morning. Dennis, a professional musician, music teacher, and former professor of music at Warner Pacific University in Portland, Oregon, found Larry, a professor of neuroscience at the Oregon Health & Science University School of Medicine, first on stage (fully clothed) at his "Music and the Brain" lectures, and then in the locker room at 6:00 a.m. Through many subsequent discussions in that locker room, we bonded over our shared love of history, literature, science, and everything there is to know and learn about music.

This unlikely beginning led us to where we are, presenting you with *Every Brain Needs Music*, our book examining how the human brain interacts with music through wholehearted listening, singing, and—most effectively—learning to play a musical instrument.

So just who are we?

DENNIS

Saying that I love music is an understatement. At age three, I was a serious listener begging for piano lessons. I began marimba lessons at six, added piano at age eight, and added trumpet at ten, and I continued to study, practice, and play all three instruments throughout my youth. In college, I majored in the pipe organ. I eventually realized that helping others understand and perform music (especially classical and jazz idioms) was just as rewarding as participating in stage, television, and radio performances. I became a professor at Warner Pacific University, where I loved teaching aural skills and piano and directed jazz ensembles, including singers and instrumentalists, for over thirty-seven years. In short, I *love* making music, teaching music, playing music, and performing on stage for audiences.

LARRY

This book is a natural extension of the talks on the neuroscience of music I have been giving for more than a decade, as well as of my own laboratory research. As a professor at the Oregon Health & Science University, I study ways to repair brain damage in people with diseases like multiple sclerosis and Alzheimer's. In particular, I study how processes like myelination, neurogenesis, and the functioning of neurons, topics covered in this book, are regulated in development and disease.

Growing up in La Jolla, California, I loved to tinker at the piano from an early age. At six, my parents took me to see a live performance of *Brigadoon* at Balboa Park's Starlight Bowl in San Diego, a fantastic outdoor venue for musical theater. When we returned home, I went directly to the piano and started playing the melodies from memory. I took piano lessons for years until my teacher, frustrated with my tendency to play by ear instead of reading assigned sheet music, advised me to "go out and play in a band or something." I took this advice, in rock and blues groups off and on through my adolescence and adult life. I eventually combined my passions for neuroscience and music to develop a talk, and now—with Dennis—a book, on music and the brain.

■ ■ ■

We hope that the interweaving of music experiences, music teaching, and neuroscience will offer readers unique insights into music, the art of teaching, learning, and creativity. Rather than presenting an extensive review of musicology or of the neuroscience of music, we hope that *Every Brain Needs Music* will serve as an introduction to the big questions in these fields and will be of interest to anyone who loves music and is interested in gaining insights into how we create music, teach and learn music, and perform and listen to music. However, you may apply the discussions in all the following movements to how humans teach and learn in general and to the underlying nature of creativity itself. Finally, we hope that this book shows why you should never be afraid to have meaningful conversations with someone when you're standing, naked, in a locker room at six in the morning.

OVERTURE

THIS BOOK involves the collision of two worlds that cannot live without each other: the world of the human brain and the world of the music that the brain creates, performs, and enjoys. It is a tale of the nerve cells and circuits that allow performers from Pharrel to Prokofiev to compose and perform their musical creations. It is also the tale of every music student and teacher seeking to perfect how the student plays an instrument or sings. It is the story of listening to a song by Elis Regina or Scott Joplin that makes your body involuntarily move to the beat, or to an Amália Rodrigues or Alison Krauss song that makes you cry. By the end of this journey, you will better understand how human beings create, practice, perform, and listen to music. You will also gain insights into how the neuroscience underlying these activities can help you appreciate the origins of your own creativity, inspire approaches to teaching and learning, and reveal whole new ways to appreciate the music and other art around you.

Before we began writing this book, we developed a survey that was sent to over one hundred composers, musicians, and music lovers to try to understand the many ways people experience and engage with music (appendix A). The rich and enlightening responses to our survey guided the writing of each movement. Dennis examined the answers through the lens of his experience as a teacher, performer, and lifelong student of music. Larry wrote about the processes and changes that occur in our brains when we compose, practice, perform, and listen to music. So the voices guiding you in the movements that follow are a combination of these survey responses and Dennis's and Larry's experiences, thoughts, and critical reviews of relevant literature.

The journey starts in the first movement, where we explore what music is, the origins of music, the reasons for music's existence, and the ways that humans have used music since the dawn of humanity.

Next, the second movement looks at how the human brain creates music. We focus on the processes of improvisation and composing, discussing their similarities and differences. We examine how areas of the brain that regulate functions ranging from empathy and impulse control to basic decision-making play major roles in music making. We also explore how the brain can create music alone or in a group of brains, and which of the brain's areas must communicate with one another when creating music, be it for a piano or a Chinese zheng.

The next three movements focus on how we teach music, how we learn to play a musical instrument or sing, and the many circuits in our brains that must communicate with one another as we practice. We follow the path from the light reflected off a page of written music to the parts of the brain that interpret what was written on that page, then to the parts of the brain that control our movements, and finally to playing or singing a note. We discuss the characteristics of a good music learner and a good music teacher, and how a motivated music learner literally rewires their brain when learning to play an instrument or to sing.

In the sixth movement we look at how the effort invested in practicing music culminates in a performance. We discuss how performing music for oneself or for an audience differs from the processes of learning to play and practicing. We address how the brain reacts to the environment where music is performed and what happens in the brain when someone experiences stage fright. We also explore the relationship between musicians and their audience, and how their brains interact with one another to influence a performance.

We dedicate the seventh and eighth movements to how the brain listens to music and how the brain comes to like or dislike different types of music. We explore how a source of music—whether instrument, voice, or recording—alters the movements of air molecules that enter the ear and stimulate specialized nerve cells that travel to many different parts of the brain, helping us perceive what we call music. We also explore how these vibrating air molecules can have powerful effects on our emotions and our sense of well-being, finally returning to the question of why music exists.

We wrote *Every Brain Needs Music* to inspire you to think about how music, like other forms of art, comes to be created and perceived in the human brain. We hope it helps you consider how challenging your brain to learn something new that involves movement, sensation, and cognition can lead to remarkable changes not only in the structure of your brain but also in who you are as a human being. And now, in the words of Mack David and Jerry Livingston, "On with the show: this is it!"

EVERY BRAIN NEEDS MUSIC

First Movement

WHAT IS MUSIC, AND WHY DOES IT EXIST?

IF AN EXTRATERRESTRIAL ALIEN landed in your backyard and (assuming that the alien somehow learned a language that you know) asked you to describe music using words, what would you say? Robert Fripp, guitarist, composer, and member of the progressive rock band King Crimson, once said, "Music is the wine that fills the cup of silence." That sounds lovely, but it is far too poetic for neuroscientists (like Larry) and neuroscience-curious musicians and musical academics (like Dennis) doing a deep dive into what music is to the human brain. For the purpose of this book, we will define music more along the lines of pioneering rock musician Frank Zappa, who declared that "a composer is a guy who goes around forcing his will on unsuspecting air molecules often with the assistance of unsuspecting musicians."

Indeed, the music generated in the mind of any human and performed by that human or by others (either live or through a recording) leads to a series of changes in the vibrations of molecules that travel through the air at around 343 meters per second (about 1,235 kilometers per hour or 767 miles per hour, to compare the speed of sound to how fast you drive a car). So, in one sense, music can be defined as the changes experienced by air molecules over time when singers sing or instrumentalists play their instruments. But if all of these changes in air molecules occur in a forest and there is nothing there capable of perceiving the patterns of these molecules as music, is there music?

The changing properties of these air molecules must somehow be detected and interpreted as music. But how? The answer is the human brain. Our brains are who we are, functioning as ever-changing computers that allow us to perceive the world around us, respond to what we perceive, and imagine and create art, literature, and music that the brains of others can perceive. For example, the brain of a composer who has spent many hours creating new music passes that information on to the brains of musicians who spend many hours practicing and interpreting that music, eventually performing that music for the brains of an audience. Each step of the way, these experiences result in both short- and long-term changes to those brains. Thus, beyond altering the behaviors of air molecules, to understand what music is we must understand how our brains perceive and respond to it.

Throughout this book, we will look at how different areas in the brain change by creating, practicing, performing, and listening to music. Distinct brain areas comprise different collections of cells that interact with one another within a given area but that also form connections with other parts of the brain, in turn forming networks of circuits that influence learning, memory, movement, sensation, and cognition.

At first glance, the human brain is a structure with many bumps (called *gyri*) and grooves (called *sulci*), some of which divide the brain into four major lobes (figure 1.1). The *frontal lobe* is where "executive functions" occur, including emotional regulation, planning, reasoning, and problem solving. It also controls voluntary movement and activity. The *parietal lobe* processes information about our sensations of the world, including temperature, taste, touch, and movement, while the *occipital lobe* is primarily responsible for vision. The *temporal lobe* processes memories, integrating them with sensations of taste, sound,

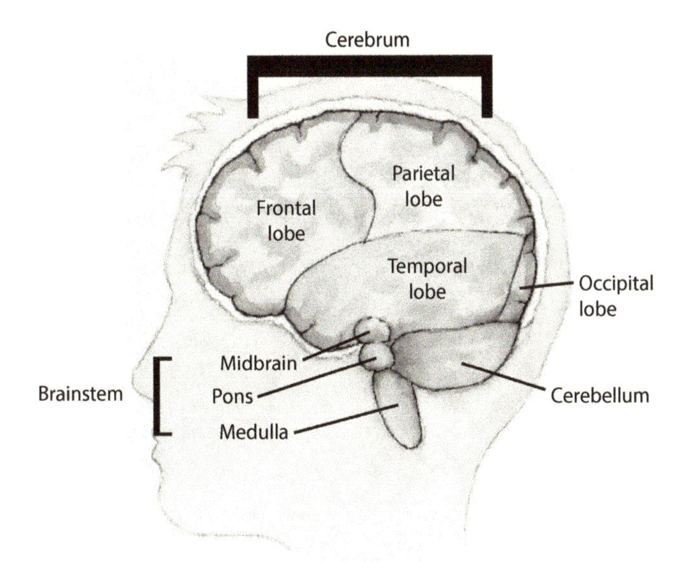

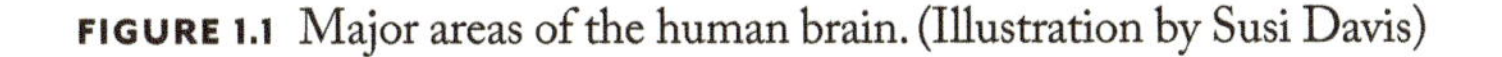

FIGURE 1.1 Major areas of the human brain. (Illustration by Susi Davis)

sight, and touch. It also contains the *auditory cortex*, which processes sound. Other major brain areas include the *brainstem*, which has many functions and connects the brain to the spinal cord, and the *cerebellum*, which coordinates voluntary movements such as posture, balance, coordination, and some aspects of speech and sound processing, including music.

These major brain structures, and many smaller structures within them, function differently from and in conjunction with one another to perceive and react to the many changes imposed on air molecules over time by music, allowing our brains to understand that what we are hearing *is* music. These structures also allow us to compose, learn to play, practice, and perform music.

NEURONS: THE ELECTRICAL UNITS OF THE NERVOUS SYSTEM

Weighing in at roughly 1.5 kilograms (3.3 pounds), your brain contains approximately 171 billion cells.[1] Around 86 billion of those cells are *neurons*—the major electrical units of our nervous systems. Neurons form the circuits that allow different brain areas to perform their functions and to communicate with one another. Neurons respond to the world around us, including to music.

Neurons have a unique structure compared with other cells. As shown in figure 1.2, this structure includes the *cell body*, which includes many tiny cellular parts that allow neurons to use energy and perform their various functions, as well as the cell *nucleus*, where the cell's DNA (deoxyribonucleic acid) is found, encoding everything cells need to function.

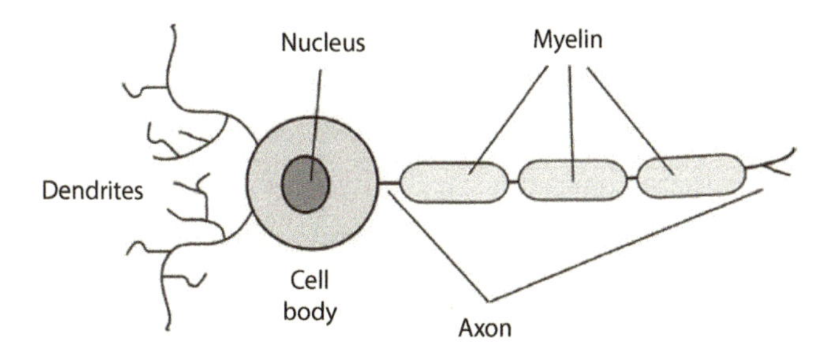

FIGURE 1.2 Structure of a typical neuron.

Extending from the cell body are branched structures called *dendrites*. The dendrites on any given neuron may be short or long, simple or highly branched, depending on where they are and what they need to connect with elsewhere in the nervous system (figure 1.2). Electrical signals enter neurons through dendrites. The signals are then passed along through an *axon*, a special branch that is also connected to the cell body; the axon is often surrounded by a substance called *myelin* (more on that in later movements). So, for any given neuron, nerve impulses coming from other neurons travel along the dendrites, then through the cell body, and then along the axon, which transmits the impulses to the dendrites of other neurons.

NEURONS TALK TO ONE ANOTHER, FORMING THE CIRCUITS OF OUR NERVOUS SYSTEMS

These connections between the axons of one neuron and the dendrites of another form the circuits of our brain, allowing us to do all the wonderful things that make us human, including creating, practicing, performing, and listening to music. A single neuron may have connections with thousands of other neurons, integrating signals from diverse regions of the nervous system. These connections, called *synapses*, might be between one axon and one dendrite, between one axon and the cell body of another neuron, or even between one axon and another. It is estimated that the adult human brain contains as many as 0.164 quadrillion (164,000,000,000,000) synapses.[2] Learning things like playing a new instrument helps form new synapses and strengthen or reorganize old ones.

Synapses may be electrical, in which case there is a direct connection between neurons, or chemical, where substances called *neurotransmitters* are passed from one neuron to another, causing the signal to continue to the next neuron in the circuit (figure 1.3). Interestingly, there are numerous different types of neurons. Some, when they fire, turn "on" parts of circuits, while

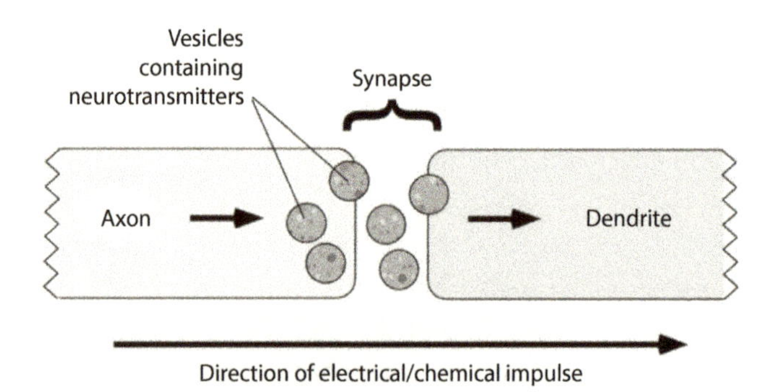

FIGURE 1.3 A chemical synapse: An axon that has "fired" its impulse releases vesicles (little hollow spheres) filled with neurotransmitters taken up by dendrites, causing them to "fire" or, if the neurotransmitter is inhibitory, preventing them from firing.

others turn "off" parts of circuits, depending on which neurotransmitter is released at the synapse. The combination of "on" and "off" chemical signals regulates how brain circuits function.

HOW WE LISTEN TO MUSIC: CONVERTING VIBRATING AIR MOLECULES INTO WHAT WE PERCEIVE AS SOUND

The perception of changes in the properties of air molecules induced by music involves a neural circuit that starts with neurons that form synapses with the inner ear's vibration sensors and ends with neurons in the *auditory cortex*, a part of the brain's temporal lobe. To appreciate this "listening" circuit (and others discussed later), you can trace the pathways outlined in accompanying figures. For this circuit, follow figure 1.4 as you read the descriptions. Keep in mind that each step is represented by a different neuron (actually, groups of neurons) and that the axons of each neuron in the circuit form synapses with the next neuron in the circuit.

The vibrating air molecules generated by musical performances or other sources of sounds enter the human ear and hit the *tympanic membrane* (also called the ear drum), which, through a group of very tiny bones, transmits its vibrations to a snail-shaped structure called the *cochlea* (Greek for snail). The cochlea is a fluid-filled structure lined with cells that have tiny hairs (called *stereocilia*).

These hair cells are vibration sensors that transmit the vibrations they receive to about 30,000 neurons whose axons form a

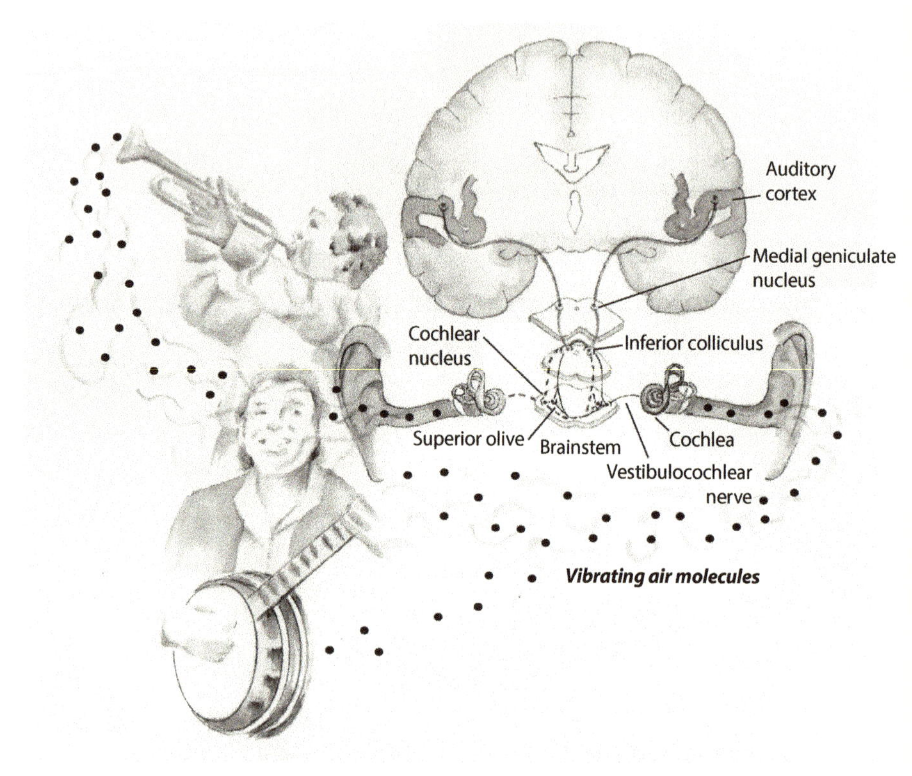

FIGURE 1.4 How sound is processed by the brain. Vibrating air molecules generated, for example, during musical performance travel through the air, are converted into nerve impulses in the cochlea, and then travel from one neuron to another, eventually connecting to the auditory cortex. (Illustration by Susi Davis)

cable called the cochlear (or auditory) nerve, which merges with another cable of axons to form the *vestibulocochlear nerve*.

The auditory axons in this nerve form synapses with groups of neurons in the brainstem in a place called the *cochlear nucleus*. Here, a nucleus means a collection of neurons all together in one place. From the neurons in the cochlear nucleus, axons either cross over to the other side of the brain or synapse with neurons in the nearby *superior olive* (which is not to say it's a *better* olive, but rather an olive-shaped structure *above* another olive-shaped structure, the inferior olive).

Next, axons from superior olive neurons synapse with neurons in a structure farther up the brainstem called the *inferior colliculus*, which, you may have guessed, is below the superior colliculus, or bump, whose neurons send axons to a very important brain structure called the *thalamus*. The thalamus is located just above the brainstem between the cerebrum and the midbrain and serves as a relay station for nearly all sensory information coming into the brain (figure 1.1). In particular, auditory information is carried by a specific part of the thalamus called the *medial geniculate nucleus* (figure 1.4). From there, neurons send their axons to the auditory cortex, where different parts of sound are sorted out. We will discuss the functions of these different areas further in the seventh movement.

Neurons from the auditory cortex transmit their signals to other parts of the brain that interpret these sounds (including the frontal and parietal lobes), giving us the experience of what we are listening to. In the case of music, that includes all its main components, including

- *rhythm*, how groups of notes form patterns that get you to clap or tap your feet;
- *pitch*, the notes being sung or played;
- *tempo*, the speed of the music;
- *contour*, how the music moves up or down;

- *timbre*, that quality that distinguishes different instruments or voices;
- *loudness*, whether the music is soft or loud; and
- *reverberation*, the way sound bounces off of surfaces in a space.

MEET YOUR MUSIC-LOVING NEURONS

It is unclear precisely how the brain recognizes and responds to music and other sounds. Why, for example, do we immediately recognize music as music after hearing just a few notes or even a simple chord, while when we hear a toilet flushing, we understand that the crescendo of rushing water is not music but rather a sign of functional plumbing?

It turns out that some parts of the human brain may respond to specific categories of sounds, including music. Many studies, relying largely on patients with brain injuries due to trauma or disease, have crudely mapped areas of the brain that process different parts of music but that have overlapping functions that also process other categories of sound. Over the past decade, more refined methods for localizing different brain functions have been developed.

One of these methods is functional magnetic resonance imaging (fMRI). An fMRI scan uses the same equipment as the MRIs that are typically used to diagnose an injury or illness. The difference is that brain fMRIs are measuring the activity of the brain, often in response to stimuli, like recorded music or an instrument being played. How do they do that?

The fMRI scan detects the changes in the flow of blood through tiny blood vessels in different regions of the brain and changes in blood oxygenation, which represent changes in neuronal activity (figure 1.5). The machines are very noisy, and for brain scans you have to lie on your back, entirely

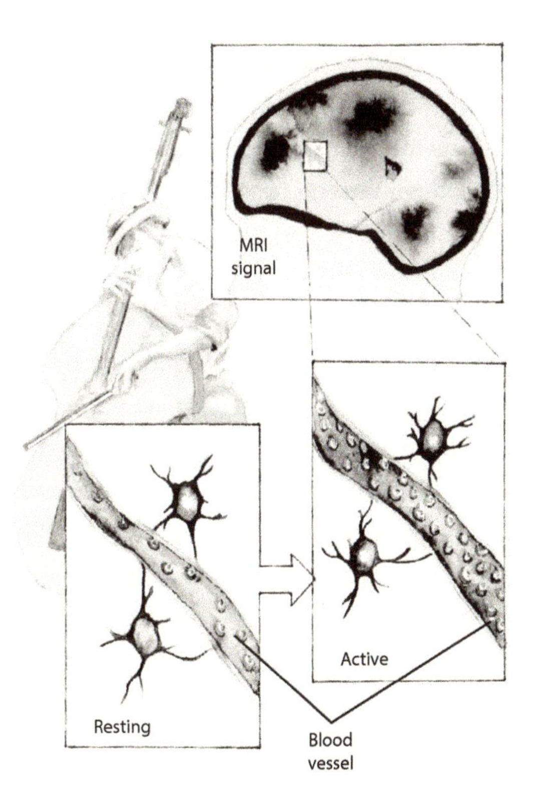

FIGURE 1.5 fMRI scans detect changes in blood flow and blood oxygenation in areas where neurons are active. Signals from neurons and cells called astrocytes can influence blood flow. The degrees of change can be color-coded and mapped to brain areas, reflecting where activity increases or decreases. (Illustration by Susi Davis)

still, with your head enclosed in a small, loud space. So, not exactly the best conditions for experiencing music, especially if you have a fear of small spaces! However, noise-canceling headphones can help while in these scanners, and it is then possible to map the areas in the brain where the flow and oxygenation of blood have changed.

Although there have been significant advances in recent years regarding how fMRI studies are performed and analyzed, there are still questions about the degree to which fMRI signals fully reflect the activities of neurons, and many variables can influence how the brains of subjects undergoing fMRIs respond to stimuli, including music. These variables include differences in how humans perform tasks, their moods, their attention levels, their levels of motivation, and other factors.[3] Nonetheless, many fMRI studies have reproducible outcomes and have been validated in animal studies. Thus fMRI studies must be interpreted with some caution. In the absence of additional data that validate a particular study, findings should be considered preliminary.

One such remarkable study examined the possibility that different types of sounds influence the activity of specific parts of the auditory cortex, the part of the brain that initially responds to sounds. Subjects lying in a horizontal position and staying completely still listened to an assortment of sounds in random order. For example, they heard someone brushing their teeth, a woman speaking, a vacuum, a car horn, a short sample of a rock song, a guitar, and, yes, a toilet flushing. Different sounds were broken into categories like language, background noise, and music.

Astoundingly, the study demonstrated that music, but not other sounds, activated specific areas in the auditory cortex in both hemispheres.[4] So, what does that mean? We can conclude from this study that specific sets of neurons may become activated in response to music, a finding that will need to be further explored in future studies. If this finding is true, then we may need to redefine music as changes in air molecules over time that induce activity in music-responding neurons in the auditory cortex of the human brain.

The findings from this study raise a number of questions. For one, it is unclear *why* these cells respond to music and what happens after they become activated. Were these particular neurons somehow prone to responding to music when they came into existence, or did exposure to music somehow stimulate these neurons to become music-specific? Another open question is where these cells project their axons and dendrites. For example, when these cells become activated, do they give a heads-up to the rest of the brain that the sounds entering the ear are music and that these signals need to be converted and integrated into all the different parts of music? Future studies will reveal the nature of these remarkable cells.

MUSIC HAS LIKELY BEEN PART OF THE HUMAN EXPERIENCE FROM THE BEGINNING

Regardless of when these music cells first appeared, the notion that such cells exist raises the possibility that the human brain evolved to engage in music. Modern humans (*Homo sapiens*) have existed for around 300,000 years.[5] The oldest known musical instruments are flutes made from bones, discovered in southwestern Germany dating to 40,000 years ago.[6] It has been suggested that Neanderthals may also have engaged in music, including instrumental music. Indeed, a fragment of a cave bear femur dated to 43,000 years ago was discovered in a Neanderthal cave in Slovenia with regular holes on one side of the bone.[7] Whether this bone was indeed a flute remains the subject of substantial debate. On the one hand, data from a study of the bone suggest that the holes were generated by scavenging spotted hyenas.[8] However, the alignment of the holes, the lack of damage on the other side of the bone, and the finding that models of the bone when played generate a diatonic scale raise doubts that the holes were made by predators or scavengers.[9] So we may not be the only hominids to have engaged in early forms of music.

The flutes found in the caves of *Homo sapiens* are relatively sophisticated, suggesting that the technology to make them occurred well before these examples were made. Flutes are also relatively advanced instruments, suggesting that they developed from more primitive instruments or musical practices. Indeed, vocal music, rhythmic music (including drumming), and other earlier forms of music were likely around much longer than the bone flutes found in Germany. Earlier flutes or other instruments may have existed, but if they were constructed from less sturdy materials, such as bamboo, they would not have survived over time. So it's possible that humans have had a relationship with music, or something approximating music, since the dawn of human existence. This notion is consistent with the discovery of cell networks in the auditory cortex that specifically respond to music, and that these networks have been part of the human brain for most if not all of our history.

WHY DO WE LISTEN, PLAY, PRACTICE, AND PERFORM?

There is little doubt that music is a core human experience. One might even say it is a human trait. Ethnomusicologist Bruno Nettl says, "All societies have vocal music. Virtually all have instruments of some sort, although a few tribal societies may not, but even they have some kind of percussion. Vocal music is carried out by both men and women, although singing together in octaves is not a cultural universal, perhaps for social reasons. All societies have at least some music that conforms to a meter or contains a pulse."[10] Thus, although the forms may differ, music is something that all groups of humans around the planet engage in to one degree or another.

But why do humans spend so much of their time and energy learning, making, and listening to music? While pondering the question, "why music?" we realized there could be an endless number of answers. As we mentioned in the overture, we developed a survey that we sent to more than a hundred musicians, composers, and lovers of music. Their many answers to the question "why music?" included the following:

- "Making music gives me a unique feeling of joy and connectedness. I feel connected to my body as the instrument, my community, and the larger universe. It's a powerful, nonintellectual experience that can transport me. My empathy is enhanced." (Joellen Sweeney, singer)
- "When I can really surrender myself to it, music connects me to my emotional self in a way that nothing else can.

It's the intellect and emotional living in one container." (Susanna Mars, actress and singer)

- "Of all the arts, it's the one that moves me—literally and figuratively. Music makes both my physical and emotional bodies move. Whether by oneself, or in a group of people, the feeling of making music has the potential to alter one's way of being. Music is a universal language that communicates more easily and deeply than words. For me, it is the easiest way to create connection between myself and others. Listening to the music of others has often led me to places that I would have never found on my own. Musicians who are game-changers/rule breakers have helped me to expand my own creative awareness and ideas." (Valerie Day, singer in the group Nu Shooz)
- "It readily affords me the opportunity to explore, discover, and express my 'inner-self.' This is a very intimate and personal process." (Bob Altig, pianist)
- "Music is connection, culture, tradition, rebellion, and expression all wrapped together in both a shared and individual experience. Music provides identity, both individual and corporate. Likely at the core of all this is a need to belong to and be a part of something larger." (Brad Tripp, music lover)

These responses reveal how ambiguous music is, how deeply meaningful it is at a personal level, and also how it serves as an organizing cultural activity. These ideas are consistent with leading evolutionary hypotheses regarding the functions of music, and they support the idea that music is an "embodied language" that promotes group cohesion by engaging specific brain areas.[11] Indeed, Charles Darwin suggested that music is a social instinct that governs our social evolution.[12] Music can promote human well-being by facilitating human contact, expressing human meaning (including storytelling), and expressing human imagination.[13] These features of music make it an excellent vector to promote social behaviors that benefit human existence.

Leonard Bernstein also explored the question "why music?" in his 1973 lecture series, "The Unanswered Question," a title borrowed from a composition by Charles Ives. In these lectures, Bernstein attempts to extrapolate the concepts regarding language put forward by Noam Chomsky to the human experience of music. Chomsky suggests that all humans possess an innate grammatical competence related to spoken language. Using numerous musical examples ranging from Hindu ragas to the symphonies of Aaron Copland, Bernstein argues that music similarly has roots in a universal language central to all artistic creation.[14] But, as the title would suggest, the question remains unanswered.

HOW DO WE USE MUSIC?

Another way to approach the question "why music?" is to consider how humans use it. Music is among the first things that infants hear because human mothers instinctively sing to their babies to calm them (granted, some sing better than others, but they all do it). It is reasonable to assume that the first human mothers similarly made music-like vocalizations. Upon birth we are dependent on others to gradually nurture and teach us to communicate through emotional interaction using both facial expressions and sounds. We hear our mothers and others sing, speak, and express themselves with rhythm, intonation, and other verbal structures (prosody) that are built into our language. Infant-directed speech "involves exaggerated vowels, repetition, higher pitches, a greater range of pitch, and a relatively slow tempo."[15] More than words, interactive pitch fluctuations help make clear the feelings experienced when attempting to teach and guide the child. "Oh," for example, can mean more than one designation depending on how it is inflected: "Oh! What is that giant growth on your nose?" (surprise) versus "Oh, you don't like my hat." (sadness) versus "Oh! I finally understand why some people like Bartok!" (excitement).

As social beings, humans want to communicate and connect with others for meaning and purpose. Iain McGilchrist states, "The 'syntax' of music is simpler, less highly evolved, than that of language, suggesting an earlier origin. More importantly, observation of the development of language in children confirms that the musical aspects of language do indeed come first. Intonation, phrasing, and rhythm develop first; syntax and vocabulary come only later."[16] It is possible, then, that our musical brains play an important role in developing other skills as our brains develop, including our strategies for expressing emotion and desire, and, later, language. Music and language come to occupy different but sometimes overlapping circuits in our brains. Our musicality can enhance the meaning of our words and our ability to express ourselves in subtle ways as we engage in speech and also in more integrated ways as we create, perform, and listen to music, by itself or paired with lyrics.

While vocalizations resembling music may precede language in the developing human brain, that does not necessarily mean that such vocalizations preceded language over the course of human history. If that were the case, one might expect to find groups of humans whose language is more music than spoken word. As it turns out, such examples can be found in modern extant tribes in the Amazon basin, including the Pirahã. The language of the Pirahã is highly musical, so much so that they can dispense with vowels and consonants and instead sing, hum, or whistle to make themselves understood.[17]

So, one possible answer to "why music?" could be that it is an excellent mechanism of communication. A great example can be found in how many cultures teach their alphabets to children. For example, in English-speaking countries (and several non-English-speaking countries), the alphabet is taught using the "ABC" song, whose melody is the same as "Twinkle, Twinkle, Little Star" and "Baa, Baa, Black Sheep." This melody is actually from an eighteenth-century French jingle by an unknown composer entitled "*Ah! Vous dirai-je, maman*," which was popularized by Wolfgang Amadeus Mozart in a piano composition he called "Twelve Variations on *Ah vous dirai-je, Maman*." When the alphabet is paired with this popular tune, children learn their letters far more quickly than they would without music.

Another example comes from the music of the Underground Railroad during the early to mid-nineteenth century in the United States. The Underground Railroad was a network of secret routes and safe houses used by escaping slaves to get to free states and to Canada. The majority of slaves were not allowed to learn how to read or write, so, since the routes were often complicated, they used songs to communicate messages and directions, and even to warn of dangers along the way. Harriet Tubman, perhaps the most famous Underground Railroad "conductor," facilitated the escape of groups of slaves using music as code for specific dangers. For example, Tubman used "Wade in the Water" to tell slaves to get into the water along a particular route to avoid being seen:

Wade in the Water.
God's gonna trouble the water.
Who are those children all dressed in red?
God's gonna trouble the water.
Must be the ones that Moses led.
God's gonna trouble the water.
Who are those children all dressed in white?
God's gonna trouble the water.
Must be the ones of the Israelites.
God's gonna trouble the water.
Who are those children all dressed in blue?
God's gonna trouble the water.
Must be the ones that made it through.
God's gonna trouble the water.[18]

In many such examples, Moses was code for Tubman herself, leading her people to freedom.[19] This is an excellent example of how pairing language with music supports more efficient and reliable information recall than language alone.

Music can inspire so much more than these practical functions in the human brain, as indicated by the answers to our survey. It may be that music provides a nexus of the known with the unknown, that it includes wonder, creativity, intellect, process, product, and a tool to get through the day. Music allows humans to say something with personal interpretation, without having to project their egos. The unanswered question therefore actually has many answers. Music exists because our brains are exquisitely tuned to generate and respond to music, to facilitate communication and interconnectedness between humans, and to arouse powerful emotional responses in musical listeners, composers, and performers.

Second Movement

HOW YOUR BRAIN COMPOSES MUSIC

HOW DOES someone come to initiate the sounds that we call music—or, as it's generally known, to "compose"? At some time in the distant past, someone, somewhere, had an idea in their mind that set the stage for a song. This may have started with a series of vocalizations that were pleasant and therefore often repeated. It may have been inspired by vocalizations that were later accompanied by banging a stick on a rock or pounding an early drum (and even later, blowing on a flute like the ones described in the last movement).

So experimenting and noodling may be the exact processes that led to the first musical composition. *Noodling*, by the way, is a technical term in the world of music for fooling around (not be confused with canoodling, used to describe another kind of fooling around—or with another form of noodling that involves the time-honored tradition of catching catfish with your bare hands).

COMPOSING AND IMPROVISING ARE DISTINCT CREATIVE PROCESSES

While noodling and other processes that occur in the brain during composition and improvisation are likely to be shared, there are significant differences in what people do when they compose compared to when they improvise.

Composing occurs by thinking in sound. The brain activates auditory imagery. By experimenting with sounds and internally "hearing" where those sounds might go, the composer hears the trajectory of the possible melody, chord progression, rhythm, and textures by combining voices, instruments, or both. Some composers conceive of whole pieces from beginning to end. Others find out what goes where as they continue to expand on the initial idea. Some work from sketches, then fill in the details. In any case, composing requires audiation, an inner hearing of music without the external performance. This could be considered the aural equivalent of an interior designer imagining a room's new decorations without yet seeing the finished room. For musicians, audiation is the foundation for performing, listening, and certainly composing.

An important distinction between composing and improvising is that when composing, material can be edited and re-edited. In contrast, improvising is unconstrained. Improvisation literally means "without provision." It's the act of letting go and spontaneously producing music. Those who engage in this seemingly unpredictable activity might analyze and tinker with the original afterwards, but the actual process of improvisation requires acceptance of what is being played or sung in the moment.

That said, improvising can only occur because it is based on structure or agreement, such as a selected melody, scale, chord progression, musical style, or an agreement to play with total freedom. A soloist with an orchestra may be directed to riff off a sketch from the score yet is confined to the orchestral prompts. One may also improvise in the style of a selected composer. In a melodic improvisation, musicians generally adhere to the melody while freely stylizing, understanding that each time the tune is played or sung the stylizing might change. Popular jazz players and singers do their thing based on the tune without violating the given melody while expressively adding their personal response to the original stimulus. A composition may be the stimulus for improvising, and the freedom of improvising may become the stimulus for a composition.

In contrast to improvisation, when our brains engage in composing, we are creating something that is intended to last and to be reproduced through live or recorded performances. To generate new music, composers often engage in both rational and emotional planning as well as their intuition. Some composers create well in casual settings, others in more formal settings. Songwriters, for example, may emphasize feelings and personal experiences. Writing instrumental music is often more complex, offering a wide variety of intellectual challenges. Hence, there may be more of an emphasis on technical matters as opposed to trying to express an emotion. Often a work may seem to the composer to have been "discovered" or "downloaded" rather than created, the composer simply being the intermediary. The piece "writes itself."

WHY DO WE WRITE MUSIC?

People are motivated to compose for different reasons than when they improvise. Composers write for certain players or singers, or to add to a specific repertoire of literature. They may write for teaching purposes, to enhance a text or dramatic work, to promote ideologies, or to sell products. Communication exists as a purpose involving the composer and those expected to hear the music (even if the composer is the only one meant to hear it).

To learn more about people's approach and attitudes about creating music, from simple songs to longer major works, we asked the music enthusiasts who answered our survey:

- How do humans compose?
- What brings inspiration? What motivates people to create?
- What are their concerns? Their intentions?
- When do people tend to compose?
- How do they go about it? What processes do composers use, and in what order do they work—text, melody, harmony, rhythm?
- What other factors—philosophical, economic, and so on—influence composing?
- When creating music, how does vulnerability and wondering how the work will be received play into a person's consciousness?

Responses to these queries suggested that most people experiment and test sounds when alone and unhindered.

Composer Tom Johnson wrote, "I sit at the piano and tinker. It is relaxing. For singing, text comes first, then melody. For instruments, it is tinkering with chords on the piano." Our respondents typically said that they might approach composing with a preexisting idea or find one along the way. Unless commissioned to produce or meet a deadline, many report composing simply to express the sheer joy of the act itself. Some compose after an intentional period of solitude and reflection. Some compose in their heads, then solidify the piece through repetition, and some put it to paper (or to screens). Inspiration can come as an urge or from a rational thought awaiting development. Since music is first aural, it has an abstract quality to its construction. It's "out there, somewhere," waiting to be meaningfully organized. The process unfolds as experimenting and seeking continue.

Regarding the composing process and attitudes, Howard Whitaker, an accomplished veteran composer, said:

> I compose either when I have leisure or its opposite, in the form of a deadline. Time of day does not matter much to me; when I'm fully engaged and it seems to be working, I can easily lose track of whether it's day or night. I normally start with one or more very basic elements: one chord, or one progression, or a rhythmic fragment. Then I experiment (often in a mechanical, "rational" way) with those ideas, trying various permutations. In that process, I often find something that I really like and at that point I get into the actual work. Inspiration? Getting work done that I'm happy with. Motivation? Most fundamentally, I think, the normal human urge

> to tinker, to take some kind of control over something. Some people manipulate sounds, some rearrange the furniture, some mess with cars, some make outrageous tweets. Same impulse. Otherwise, often working with a great performer is a great motivator. Sometimes it's money (though my experience has been that I rarely do my best work when motivated primarily by money).

Howard represents one composer of primarily complex "art music," educated traditionally with a doctorate in composition. Coincidentally, his approach matches the approach of many other composers from a wide variety of backgrounds.

For another style, one less cerebral and more visceral in intent, Jamil Kassab describes being motivated by a recreational approach:

> I tend to compose when emotions are piling up. I've realized about myself that I compose out of sorrow or hardship more often than joy and celebration. I just pick up my guitar and start playing what I know, and from there I start expanding on the known to create something new. I start with a foundation and build upon that. Knowing it is one of the only and primary ways I can fully express myself and pour myself out without trying to formulate correct "words" to express emotions adequately. I am better with sounds than words.

Songwriter Sylvia Gray gives some insights onto how lyrics can inspire music: "I compose to poetry. I seem to need the text to motivate me, and I believe I enhance the words and give them a second life in a way. I've had poets find new meaning in their own work by the art songs I have written. I work a long time with the text, trying to find how the words should fall rhythmically and naturally. Then I work on the melody."

In contrast, Ira Gershwin put words to music written by his brother George, but would often start by placing "dummy lyrics" to the music to get the cadence before adding the final lyrics. For example, for one piece of music, Ira initially wrote, "Roly-poly; eating solely; ravioli—Better watch your diet or bust." This later became "Just go forward; Don't look backward; And you'll soon be—Winding up ahead of the game." The final version from this hit song was "I got rhythm; I got music; I got my man—Who could ask for anything more?"

CURIOSITY: A KILLER OF CATS BUT A MOTIVATOR FOR COMPOSERS

Human beings (and cats) constantly explore the world around them in search of new experiences and information. This behavior is described as curiosity, a state of information seeking that can be internally motivated.[1] Curiosity helps us to fill gaps in our knowledge that can benefit us throughout life, and it has been described as an essential feature of learning.[2] Curiosity has also been implicated as an essential component of creativity, through which existing knowledge can be transformed into something novel.[3] It is not surprising, then, that the statements by composers just quoted suggest that the state of curiosity is an essential part of the act of composing music.

No matter what process a composer uses to create new music, including noodling, curiosity is part of what drives the discovery of new themes or other musical sequences as a musical composition evolves.

Many composers incorporate certain themes or musical styles into their music that listeners can identify as being at least somewhat unique to that composer. However, these musical signatures can change over the course of a composer's lifetime. Curiosity is likely to be a driving force for such changes. One great example can be found in the piano sonatas of Ludwig van Beethoven (born 1770, died 1827). By the time Beethoven was thirteen, he had already written three piano sonatas, but these are often set aside from the following thirty-two because, many argue, when he wrote them he was barely getting started and they just aren't all that great (although we think they're still pretty good). By the time he was twenty-five, he had begun writing what would be considered his early piano sonatas, which were at first heavily influenced by the works of Haydn and Mozart. Things really started to change with Piano Sonata no. 8 (*Pathétique*, the second movement of which Billy Joel put words to in "This Night" on his best-selling album *Innocent Man*). With this sonata, Beethoven departed from Haydn's and Mozart's influences and started exploring all new themes. This departure would lead to additional explorations into new musical directions, giving rise to the fantastic music that followed, including Piano Sonata no. 14 (*Moonlight*), whose first movement is perhaps Beethoven's most famous, besides the opening bars of his *Fifth Symphony*. He wrote this sonata for the woman he wanted to marry, Giulietta Guicciardi.

Beethoven's curiosity, along with his desire to engage in canoodling, may have led to some of the noodling that gave us these early Beethoven sonatas.

Things changed again at the beginning of the middle period of Beethoven's piano sonatas. From this time forward, Beethoven very intentionally wanted to explore new directions in his compositions. We know this because after finishing Piano Sonata no. 15 he wrote to his close friend, Czech-born mandolin player and violinist Wenzel Krumpholz, to announce that "From this time forward, I'm going to take a new path." This new path started with Piano Sonata no. 16 in 1802 and included works such as the *Waldstein* (Piano Sonata no. 21) and *Appassionata* (Piano Sonata no. 23), a remarkably passionate piece of music with a very simple middle movement. Beethoven continued to explore this way through Piano Sonata no. 27 in 1814, when he decided to change paths once again.

Starting with Piano Sonata no. 28 in 1816, the late sonatas include some of Beethoven's most technically challenging music, including the *Hammerklavier*, which many believed to be unplayable (Liszt proved everyone wrong when he performed it in front of an appreciative audience). These sonatas also represent some of Beethoven's greatest exploratory behavior. They were written at a time when Beethoven was deaf and experiencing significant pain, both believed to be linked to chronic lead poisoning.[4] But his curiosity drove him to write some of the most unique music of his lifetime. His final piano sonata, Piano Sonata no. 32, whose second movement is a series of variations, includes an entire page of syncopated rhythms, predecessors to ragtime and jazz.

Beethoven's drive to create something novel is a reflection of his state of curiosity. Our brains experience a sense of reward when we create something new in the process of exploring something uncertain, such as a musical phrase that we've never played or heard before.[5] When our curiosity leads to something novel, the resulting reward brings us a sense of pleasure. A number of investigators have modeled how curiosity influences musical composition. In the case of Beethoven, computer modeling focused on the thirty-two piano sonatas written after age thirteen revealed that the musical patterns found in all of Beethoven's music decreased in later sonatas, while novel patterns, including patterns that were unique to a particular sonata, increased.[6] In other words, Beethoven's music became less predictable over time as his curiosity drove the exploration of new musical ideas. Curiosity is a powerful driver of human creativity.

IMPROVISATION ACTIVATES AND DEACTIVATES CERTAIN BRAIN NETWORKS

Given the diverse ways that composers and improvisers approach creating new music, we questioned whether the conception of music depends on distinct brain regions or if these regions differ from person to person. We have already seen that composing involves both planning and revising combined with a creative process, driven in part by curiosity, that overlaps at least somewhat with the process of improvisation. To understand how the brain functions during the act of composing or improvising, studies have examined both individual, exceptional, professional musicians and groups of musicians who compose or improvise similar styles of music.

Some of the earliest reports involved fMRI studies of musicians improvising on piano keyboards. For example, Bengstsson and colleagues[7] studied how the brain changed during improvisation in eleven male Swedish concert pianists, all right-handed, similar in age, education, and the age when they started playing. All the subjects received their master's degrees in piano performance from the Royal Academy of Music in Stockholm (they were *not* clones, but pretty close). These pianists were asked to play a small piano keyboard with one octave (F to E) using their right hand while lying down in an MRI machine where they could see instructions and music projected on a screen, all while keeping their heads perfectly still by biting on an immobile bar. Scans were compared for three different conditions: "improvise," where subjects were shown sheet music with a "template" of music and asked to improvise based on that pattern and remember what they improvised; "reproduce," where subjects replayed what they improvised from memory; and "rest," where subjects were shown a blank screen and simply relaxed (as much as you can relax with your head inside the small hole of a giant noisy machine while biting down on a bar). Some subjects were also in a "free improvisation" group, where they could improvise without worrying about memorizing what they played.

The authors found that the act of improvising in this manner, both in the "free" and "template" groups, involved the activation of a number of areas in the frontal and temporal lobes of the brain (figure 2.1), including the

- *right dorsolateral prefrontal cortex*, which is involved in how we react to stimuli, among other functions;
- *presupplementary motor area*, which has been linked to how we alter our behavior with regards to changes in sequences of events;
- *rostral portion of the dorsal premotor cortex*, which may be involved in the memory for movement during specific tasks; and

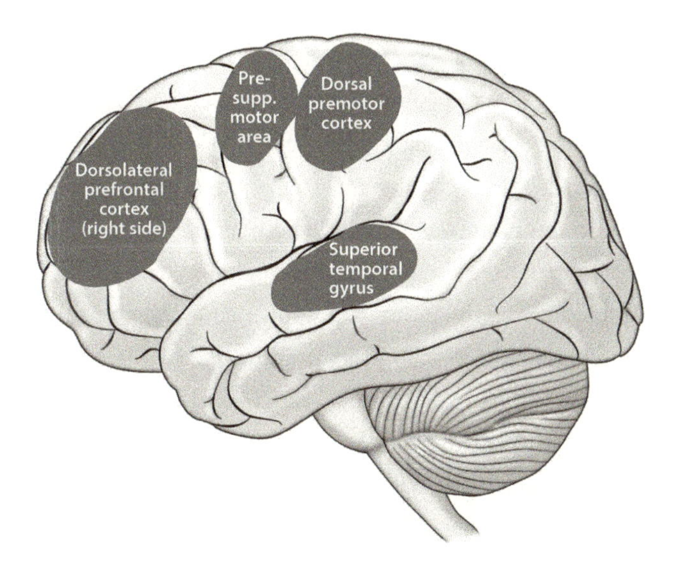

FIGURE 2.1 Areas activated in the brains of classically trained pianists who were shown sheet music with a "template" of music and asked to improvise based on that pattern and to remember what they improvised.

- *left posterior part of the superior temporal gyrus*, which includes the auditory cortex involved in processing sound as well as Wernicke's area, which is involved in language comprehension.

The authors speculated that these areas are part of a network involved in musical creation. However, we must remember that, similar to concerns about any studies involving composition or improvisation (as well as the caveats of fMRI studies mentioned in the first movement), the conditions of improvisation in this study were far from typical for most musicians (beyond the whole "lying completely still in an MRI scanner and biting on the bar" thing). Some of these areas may have been activated because the subjects were responding to the template of music, which may have triggered specific memories or associations that they might not have experienced if asked to improvise in any desired style without a musical prompt. Furthermore, the ability to improvise varies greatly between musicians, and this variation may have clouded the interpretation of the data.

In contrast to the group of Swedish classically trained pianists, Charles Limb's laboratory examined improvisation in six professional right-handed male jazz pianists between the ages of twenty-one and fifty.[8] Charles Limb, a surgeon, neuroscientist, and musician, has made substantial contributions to understanding the neural basis of musical creativity. This study's design had a number of important differences compared to the Swedish pianist study. Pianists were again in an MRI machine playing a keyboard while in a horizontal position, but they were jazz musicians, who are typically outstanding improvisers (classical pianists are not known to be outstanding improvisers, although there are exceptions). Furthermore, they were asked to perform a very different task than the Swedes. They had four conditions:

- *Scale control:* Subjects played a one-octave C major scale (playing from one C to the next highest C) in quarter notes (notes held for one beat of the music).
- *Scale improvisation:* Subjects performed a simple improvisation task in which they improvised a melody but within the same C major scale quarter notes and octave.
- *Jazz control:* Subjects were asked to memorize an original jazz composition before being tested, and then played the composition along with a prerecorded jazz quartet.
- *Jazz improvisation:* Subjects performed a complex improvisation task in which they were free to improvise to the same recorded jazz music.

Remarkably, unlike in the study with classical pianists, there was widespread *deactivation* of brain areas during both simple and complex improvisation compared to the control conditions (figure 2.2). These areas included some of the areas activated in the brains of the Swedish pianists (like the *dorsolateral prefrontal cortex*). Other areas—such as the *frontal portion of the medial prefrontal cortex* and areas involved in both sensory and motor functions—were activated in the brains of the jazz musicians. Thus spontaneous jazz improvisation by professional jazz pianists may lead to both activation and deactivation of distinct brain areas for both simple and complex music.

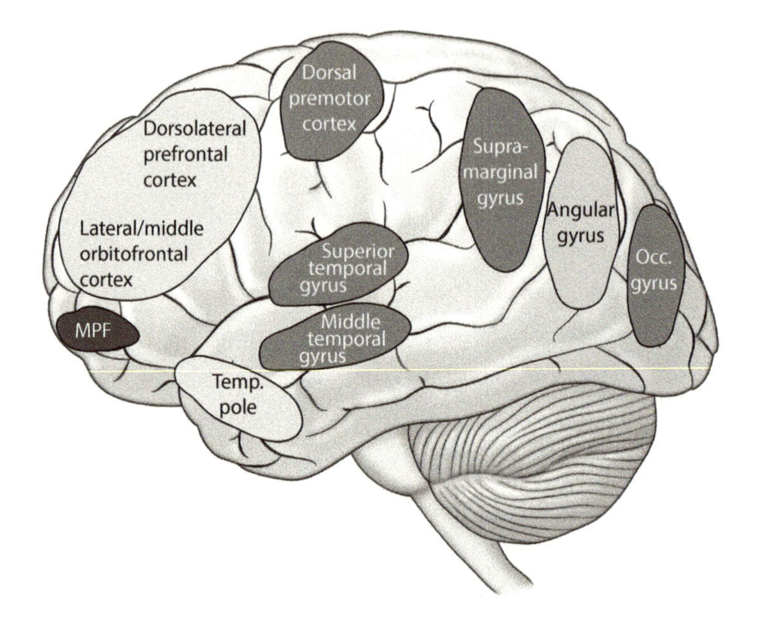

FIGURE 2.2 Areas activated in the brains of jazz pianists improvising along with a piece of recorded jazz music. Areas in light gray were deactivated while dark gray areas were activated. Note that some of the areas activated here are the same as in figure 2.1, while others that were activated in the classical music study are deactivated here (for example, the dorsolateral prefrontal cortex). MPF = medial prefrontal cortex; Occ. = occipital; Temp. = temporal.

A few of the areas that were deactivated in the solo jazz pianists are believed to encode a framework that monitors, evaluates, and corrects our goal-directed behaviors. Some people think this framework keeps us from doing things that are "outside the box." The fact that these areas become deactivated during solo improvisation suggests that we need to "let go" of that framework to create something new.

BRAINS THAT GO IT ALONE WORK DIFFERENTLY THAN BRAINS THAT WORK TOGETHER

Another study of jazz pianists from the Limb lab explored brain regions that can influence the process of improvisation in response to other musicians.[9] Jazz and other improvisational musicians often "trade" solos back and forth, sometimes for a preset length of music. For example, "trading fours" means that each musician will improvise for four bars. The interaction is often likened to a conversation, with each solo a voice responding to another.

In this study, pairs of expert jazz musicians played together, with each able to each hear what the other was playing but with one of them in the MRI machine with a keyboard and the other in the control room with another keyboard. The tasks for this experiment were similar to those of the Swedish study:

- *Scale control:* Musicians alternated playing a selected minor-sounding scale, the D Dorian scale, in quarter notes.
- *Scale improvisation:* Each musician was instructed to improvise within the Dorian scale for four bars, and then the other musician responded with his or her own improvisation (trading fours).
- *Jazz control:* Musicians alternated playing four bars of a memorized novel jazz composition.
- *Jazz improvisation:* Each musician traded fours, improvising on the same piece of novel music.

Interestingly, this musical conversation activated many areas used in verbal conversations. For example, the brain of the improvising subject in the MRI room responding to the musician in the control room demonstrated activation of the inferior frontal gyrus, which has a number of functions, including the processing of speech and language (as part of a language center called Broca's area), and activation of the posterior *superior temporal gyrus*, which is involved in auditory processing, including language. In particular, these brain regions are concerned with language syntax. In addition, the study revealed that trading fours leads to deactivation of structures involved in processing the meaning of language (the *angular gyrus* and the *supramarginal gyrus*). In other words, the brain treats these musical conversations much the way it treats verbal communication, but in a way that lets go of "meaning," in linguistic terms.

In a surprising finding, an area that was deactivated when jazz musicians improvised alone (the *dorsolateral prefrontal cortex*) was activated in musicians who were trading fours. Because this area has been linked to conscious self-monitoring of behavior, it is possible that this area needs to become activated under these circumstances because (in the case of trading fours) the musician responding to the previous four bars from their partner is consciously evaluating whether what they are playing works well in terms of melody and rhythm with what they just heard. We can interpret these findings by saying that although we are "letting go" when we improvise by ourselves, we are engaging in a much more constrained activity when we improvise off of what others have just played.

THE CLASSICAL BRAIN VERSUS THE JAZZ BRAIN

These studies suggest that improvisation requires combinations of activations and deactivations among several structures within the brain. But each study implicates different brain regions and differences in which areas become activated or deactivated, leading to some uncertainty about the significance of any given brain region. Why, for example, were the results for the Swedish pianists so different from the results for the solo jazz musicians?

First, as mentioned, the two studies were designed differently, with very different tasks (improvising off a template of sheet music in the first study versus improvising off a scale or a memorized piece of jazz music with recorded accompaniment in the second). Further, the subjects in the first study were all trained classical musicians who received graduate degrees from the same institution and who are unlikely to have had substantial experience with improvisation, while the subjects in the second study were jazz musicians with varied backgrounds and training who likely spent a great deal of their time improvising. This fact suggests that classically trained musicians utilize different processes when they improvise compared to jazz musicians, or that improvising in the style of classical music requires a different set of processes than jazz improvisation. Consistent with this idea, one study found differences in how different brain areas connected with one another in classical musicians compared with jazz musicians.[10]

To further explore this question, Barrett and colleagues studied the brain activity of a classically trained musician with exceptional improvisational skills.[11] These investigators enlisted

Venezuelan classical pianist Gabriela Montero, who is both a gifted performer of classical music and also a remarkable improviser. Montero is a unique "outlier" even among highly skilled classical pianists. She started playing the piano before she could speak and began playing concerts at age five. Her performances include playing classical repertoire as well as improvisations often suggested by audience members. Although the study examining her brain as she improvised has a sample size of exactly one subject (never a good idea when trying to make conclusions about how the human brain works in general), the authors correctly reason that the nature of expert performance is typically studied through the examination of exceptional individuals, and Ms. Montero certainly meets those criteria.

Like the Swedish and jazz pianists, Ms. Montero played a keyboard in an MRI machine but, unlike in the previous studies, she was asked to play a chromatic scale, a composition that she had learned and memorized (Johann Sebastian Bach's Minuet in G Major), and finally, improvised music (in the style of the Bach minuet). For Ms. Montero, at least, improvisation (compared to the scale and the memorized music) involved the activation of numerous areas throughout the brain, including those activated in the other two studies. While some areas also demonstrated deactivation (such as the middle temporal gyrus, the inferior parietal lobule, the angular gyrus, and parts of the cerebellum), these areas differed from the areas deactivated in the solo jazz musicians.

Thus, either Ms. Montero's brain engages in improvisation in a unique way (and there is little doubt that she has a unique brain), the way the study was done elicits different sequences of activations and deactivations in different areas of the brain compared to the jazz study, or the specific strategy Ms. Montero uses to improvise is different from the way solo jazz musicians improvise. The latter is almost certainly true. Ms. Montero improvises on pieces of music in the classical style of other composers, much like the task she engaged in for this study. So it is possible that the degree and areas of brain deactivation of an improviser may depend on the degree to which an improvisation is "outside the box" (for example, breaking the rules set by the opening bars of a piece of music). Similarly, given that certain areas (like the angular gyrus) became deactivated in both Ms. Montero's brain and in the brains of jazz musicians who were trading fours, both tasks may involve "letting go" of processes that involve linguistic meaning.

Because all the improvisation studies we have described so far have involved pianists, we might wonder whether studies of musicians playing other instruments or vocalists would have the same results. One study examined a group of jazz pianists, saxophonists, guitarists, trumpeters, trombonists, French horn players, and bassists.[12] Subjects were asked to learn and practice both vocalizing (humming or singing) and imagining vocalizing (singing in one's head) four Charlie Parker melodies from the Bebop era of jazz: "Au Privave," "Now's the Time," "Blues for Alice," and "Billie's Bounce." As with the other studies, subjects were in an MRI scanner and had to remain very still during each of these tasks:

- *Vocalize prelearned:* Subjects vocalized the pieces as they learned them.

- *Vocalize improvised:* Subjects vocalized improvised music based on each piece.
- *Imagine prelearned:* Subjects imagined vocalizing each piece.
- *Imagine improvised:* Subjects imagined improvising on each piece.

Interestingly, consistent with the studies of classical improvisation by the Swedes and Gabriela Montero, a number of brain areas, especially in the frontal lobe, became activated (including the dorsolateral prefrontal cortex) during vocal improvisation. This fact suggests that the act of improvisation when constrained to a prelearned piece of music requires a set of rules encoded by the dorsolateral prefrontal cortex. In addition, Broca's area, the language center that was active in the trading fours study, was activated in subjects engaging in both vocal and imagined vocal improvisation. It is possible that Broca's area is activated in this case because the music is vocalized even though it is in response to instrumental cues.

These authors also performed a type of analysis not done in the earlier studies. In addition to measuring the activity in specific areas of the brain, they used fMRI data to examine activity between different areas of the brain. In particular, this study examined how the strength of connections between different brain regions ("networks") changed during imagined and actual vocal improvisation. Consistent with the notion that some form of "letting go" occurs when we improvise, they found that many of these areas become "disconnected" from one another during the vocal improvisation task (for example, the strength of signals between different brain regions decreased). This finding tells us that improvisation appears to be an internally directed behavior that requires less connectivity between certain brain areas to generate new music. In other words, we turn down connections between some brain areas to "let go" and improvise.

IMPROVISING BRAINS ARE NOT ALL CREATED EQUAL

As we hinted previously, not everyone is great at improvisation. It's truly a struggle for some people. Maud Hickey from the Center for the Study of Education and the Musical Experience at Northwestern University states that free improvisation "is an improvisation that *cannot* be taught in the traditional sense, but experienced, facilitated, coached, and stimulated."[13] Similarly, composer, electronic music performer, installation artist, trombone player, and scholar George Lewis, who is an outstanding improviser, argues that the best way to get someone to improvise is to "throw them in the deep end and work with what naturally happens."[14] But what about people who seem to have a "natural talent" for improvising music (like George Lewis or Gabriela Montero)?

One small study suggests that the brains of excellent improvisers are structurally different from the brains of lesser improvisers.[15] The study examined thirty-eight students with varied musical backgrounds from Wesleyan University and the Hartt School of Music. They were instructed to improvise in response to a piece of music after listening to and briefly playing along with the music in an MRI machine. Their level of "creativity" was judged by expert jazz instructors. The authors reported that the better (more creative) improvisers had lower volumes (that is, smaller brain structures) in an area involved in the processing of visual information and the categorization of objects, the recognition of faces, and other aspects of memory (the right inferior temporal gyrus); in a part of the brain involved with learning and memory (the right and left hippocampus); and in the region activated during both singing and speaking (rolandic operculum),[16] which is associated with higher-order somatosensory processing,[17] such as the perception of pressure, pain, or warmth.

The significance of these differences is unclear, and, given the small number of subjects and the qualitative nature of how "creativity" was measured in this study, we should be cautious about reading too much into these findings. Nonetheless, if structural differences in the brains of exceptional improvisers are linked to their improvisational skills, we might ask whether they were born with these differences (through genetic or early environmental influences) or if they developed through a lifetime of engaging in improvisation.

BRAIN ACTIVITY WHEN COMPOSING MAY BE DISTINCT FROM IMPROVISING AND OTHER CREATIVE ACTIVITIES

Like studies investigating improvisation, studies exploring the neuroscience of composition have depended on groups of composers and individual composers. For example, neuroscientists

Daniel Levitin (a leader in the field of the neuroscience of music) and Scott Grafton convinced Sting (Gordon Matthew Thomas Sumner), the solo artist and former singer, songwriter, and bass player for the New Wave rock band the Police, to both listen to and compose music while being analyzed in an MRI machine.[18]

They found that when Sting was composing (imagining a new piece of music) or even imagining different parts of a composed piece (such as melody and rhythm), the changes in levels of activation in groups of brain regions were distinct from those activated by imagining new prose or visual art. Imagining melody alone and melody plus rhythm for these new compositions led to the most similar patterns of activation, while imagining rhythm alone was less similar than melody plus rhythm compared to melody alone. These findings suggest that, for Sting at least, melody drives the process of composition.

Although looking into Sting's brain might give us some idea about what our neurons are doing when we create music, additional insights have come from studying groups of composers who use similar instrumentation in a particular style of music. A study from the laboratory of Dezhong Yao at the University of Electronic Science and Technology in China examined a group of composers as they composed music for a Chinese zheng.[19] The zheng (筝; also called the guzheng 古筝 or Chinese zither) is a Chinese plucked string instrument that archeological evidence suggests has been around for approximately 2,500 years. It typically has twenty-one, twenty-five, or twenty-six strings, is 64 inches (1.6 meters) long, and is tuned in a major pentatonic scale (comprising only five notes per octave, instead of the seven notes in the heptatonic scale most commonly used in Western music). It is played with both hands, often with fingerpicks, with the performer in a sitting position. The zheng has a large, resonant soundboard made from the wood of the empress tree (*Paulownia*) and generates sounds that are reminiscent of both harps and lutes.

In their study, the Yao lab examined seventeen right-handed male and female composers from the Sichuan Conservatory of Music. They had all been composing for more than five years, though none of them had any experience with a zheng (other than being familiar with how the instrument sounds or with typical zheng music). Therefore, the investigators reasoned, the results from the study could rule out the effects of specific motor experience and memory in the compositions created by this group of composers.

The composers were examined in an MRI machine and scanned while resting (but not falling asleep). Then they were told to compose a piece of zheng music with their imagination, using a staff with a few notes as a starting point (which they could see while being scanned) (figure 2.3).

Immediately afterwards, the composers were asked to write down what they had composed to verify that the composing process had been scanned.

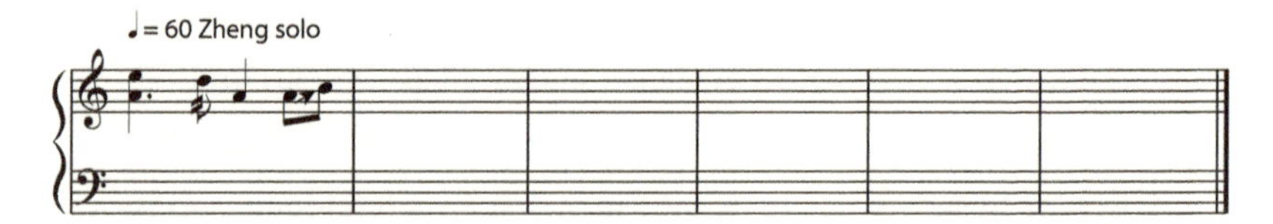

FIGURE 2.3 The prompt that subjects were shown as a starting point to compose zheng music.

Interestingly, as in the study of vocal improvisation discussed previously, this study from the Yao lab examined how the connectivity between different brain regions was altered during the act of composition. As in the study of improvising vocal music, certain areas in the brain became less connected during the act of composing compared to resting. In particular, connectivity decreased in networks involving visual and motor areas, as did an area involved in the processing of words (the so-called lingual gyrus hub, located in the occipital lobe). Given that these composers were creating music for an instrument they had never played, it is possible that these areas are not needed for imagining instrumental composing as they might be for jazz musicians using a keyboard or vocalizing music, as in the previous studies.

Interestingly, the zheng composers in this study also showed increased connectivity in other areas, including between areas implicated in empathy, impulse control, emotion, and decision-making (the anterior cingulate cortex) and areas of the so-called default mode network (including the right angular gyrus and the superior frontal gyrus). The default mode network has been implicated in a wide range of behaviors, including daydreaming, self-reflection, thinking about others, remembering past events, and planning for the future. This increased signaling between the anterior cingulate cortex and the default mode network during the act of composing may reflect the need for a specific type of planning that includes the free generation of individual notes (the melody-driven process we saw with Sting) combined with the desire to communicate specific emotional meaning in the music.

The Yao lab followed up the zheng study with a second study that sheds light on both the improvisation and composing processes.[20] This study focused on a collection of brain areas referred to as the multiple-demand system, located in the frontal and parietal cortex. This collection of areas is activated when we engage in complex cognitive tasks, including novel tasks. It has been suggested that the multiple-demand system is required for "fluid intelligence," the ability to solve new problems, use logic in new situations, and identify patterns independent of language.[21]

To explore the possibility that the acts of improvising or composing might require the multiple-demand system, the Yao lab examined a group of male and female composers, again from the Sichuan Conservatory of Music, who were studying musical composition and who were experienced piano players. These composers were scanned in the MRI system under these conditions:

- *Baseline:* Subjects were asked to think about nothing.
- *Familiar:* Subjects were asked to imagine improvising or composing off of a familiar piece of music (Ludwig van Beethoven's "Für Elise").
- *Unfamiliar:* Subjects were asked to imagine improvising or composing off of an original piece of music.

The cues for each of these tasks involved showing images of two bars of sheet music from each piece (a blank staff was used for the baseline condition). Investigators saw broad activation of the multiple-demand system during both the familiar and unfamiliar tasks. However, the unfamiliar task induced more activity in one part of the system (the intraparietal sulcus), an area involved in a number of functions, including trying to understand the intent of others.

This study suggests that composing or improvising involves the multiple-demand system for processing novel melodies but also suggests that novel cues require different activities compared to familiar cues. It is possible, therefore, that the brains of composers who generate music inspired by familiar music are using a process that is distinct from that used by composers who generate entirely new melodies.

REPRISE: BRAIN ACTIVATION DURING COMPOSITION VERSUS DURING IMPROVISATION

These findings and the responses to our survey suggest that while musicians and composers may have unique approaches to making new music, they share at least some common neural functions when engaging in these creative processes. However, the fMRI studies we reviewed suggest that the brain may utilize different pathways when creating new music, depending on whether someone is improvising or composing. In addition, the style of music, the stimulus for the creation of a piece of music, the composer's curiosity, and whether the music is simply imagined or performed (vocally or using an instrument) may result in profound differences in which brain networks are involved, or even whether the same networks or brain regions are activated or deactivated. These findings suggest that some parts of the brain engage in very different ways depending on what we might see as subtle differences in improvising or composing tasks, and that we should consider these differences when teaching, learning, or honing skills related to creating music.

Third Movement

PRACTICING MUSIC, PART I

THE PARTNERSHIP OF MOTIVATED MUSIC STUDENTS AND MOTIVATED MUSIC TEACHERS

HOW DO musicians learn to perform music? Before improvising, trading fours in a musical conversation, or performing the music created by some composer—who, as Frank Zappa said, has forced his or her will "on unsuspecting air molecules" with the assistance of "unsuspecting musicians"—those unsuspecting musicians must learn to play and, when appropriate, sing. In the next movement we will discuss the circuits in the brain that are involved with learning to play an instrument. But this movement draws largely on Dennis's experiences over several decades teaching music to many students as we explore the characteristics of effective learners and teachers. We focus on what Dennis has witnessed firsthand in students learning to play an instrument, and how he navigated the challenges of adjusting formal and prescribed methods of teaching to each student's preferred learning process. Although we focus on instrumental music, many of the principles we discuss also apply to vocal music.

Learning to play an instrument requires tightly controlled movements of the hands, fingers, and sometimes feet (for the pedals of a piano, giant pipe organ, or drum set, or to coordinate playing an instrument while marching). Connecting the physical gestures of playing an instrument with the simultaneous awareness of the sounds those movements generate stimulates a sense of cause and effect. As the stories of some of Dennis's students illustrate, the act of creating sound carries an immediate sensation, ranging from extreme joy when things go well to frustration, stress, or even horror when they do not.

TEACHING OLD DOGS NEW (MUSICAL) TRICKS

One memorable student was in his early sixties, freshly retired from teaching physical education for several decades. Since college, he had loved jazz piano. He craved the sounds, the feel, and the harmonies, but he knew nothing of what was involved in the process of performing music. He had never played a musical instrument. Dennis will never forget when this new student asked if, at his age, he could learn to play piano, read music, and ultimately improvise. "Yes," Dennis told him.

This student was highly motivated to learn. As with any subject, motivation and effort are important keys to learning—but are they enough? Dennis's student obviously sensed he needed help to achieve his long-term goals. We know that, when learning, humans need to see progress to stay motivated. Learning the strategies that help get us to our goals using a stepwise process is perhaps the best way to make progress. Furthermore, basic science research suggests that developing learning strategies, either taught or developed independently by learners, may be even more critical than motivation alone for the brain plasticity required when learning a complex task like playing a musical instrument.[1] Dennis's own teaching experiences are consistent with that idea, and he has come to understand how a teacher provides those strategies (or helps students find such strategies) to support even motivated students.

In his teaching, Dennis always shows students how to do assignments as clearly as possible. The student's role is to go over the assigned material *daily*—slowly, methodically, and mindfully. For adult learners the number one enemy seems to be lack of patience, which can diminish motivation and lead a student to give up. So part of the strategy for them is finding ways to settle in and simply be faithful to the skill-building exercises and pieces without heavy expectations of immediate success.

For Dennis's retired student, lessons began with weekly half-hour piano instruction and daily hour-long practice sessions based on concrete written assignments. Fortunately, he had a number of characteristics that helped him succeed. First, he was fairly realistic in his expectations, eventually realizing that smaller successes would lead to greater goals. Second, he was highly motivated to learn because of his love of jazz. Third, he trusted Dennis's assignments, and, knowing they were specific to him, he grew steadily. Lastly, after a few lessons, he confessed that he had thought he'd be playing piano in a few weeks until he realized that to play at a high level could take years. He could appreciate this realization in part because of his own life experience; as a tournament-level tennis player and physical education teacher, he had already experienced the need for patience in athletics. He made the connection to musical reality.

Years later this student reported that his life was altered by the music lessons and his consistent practicing. He noticed that music enhanced his analytical thinking skills and spread to other areas of his life. He was turning into a thinker and greatly enjoying his expanding outlook. His mental habits changed. His choices of entertainment became more intellectually challenging, and so did his critical thinking. Music shifted him from a more superficial, frivolous take on life toward more depth and complexity. He was happier, and he felt more rounded and complete. He had developed into a lifelong learner—and a musician.

Another adult student Dennis met already played the piano very well, and she had strong technique and reading ability, but she felt controlled by the printed music. So in midlife she decided to take improvisation lessons, hoping to loosen up her approach to music making. As we discussed in the previous movement, improvisation involves communication through specific musical content, so some knowledge about how music works is helpful. However, despite her excellent music reading ability, she resisted theory, the understanding of its ingredients and how music works. So, in creative ways, Dennis worked to help her release what she heard within herself. The strategy here was to help this student come to the realization that the long-term goal of improvisation could be achieved through stepwise achievements in understanding music theory. In fact, ten years after her two years of lessons with Dennis she decided to take theory classes at the local community college and reported that she wished she had been open to theory earlier. She was elated by how the added knowledge and understanding of music enhanced her playing. It also opened cognitive doors, which she found enlightening and, combined with playing, brought her even more joy.

The experiences of these students illustrate how, at any age, learning new skills like playing a musical instrument can have a broad range of effects on our brains. These students also illustrate how the brain is capable of remarkable plasticity.[2] And these anecdotes are supported by studies in experimental neuroscience, which have shown that although the plasticity of the brain declines with aging, it is still capable of remarkable structural changes given the right degree of challenge (discussed further in the movements that follow). In short, you *can* teach an old dog new tricks; it just takes longer.

MUSIC PRACTICE IS A LONG AND WINDING ROAD

Not all of Dennis's students found such success in their musical journeys. Dennis recalls a case of a man who seemed to have found that everything in life had come easily for him. He quickly learned and excelled at everything he tried, including getting through multiple postbaccalaureate degrees with ease. He came to Dennis for help because, he said, he wanted to express himself through music, specifically via piano.

After six months of weekly lessons that seemed to Dennis to be quite successful and joyful for this student, he was shocked when, without saying a word, the student suddenly stormed out, never to be seen again. He must have become so aggravated at himself for not easily getting his fingers to accomplish what he perceived either visually or aurally that he gave up. For this student, many elements of life had been a snap. However, it seemed that learning to play a musical instrument took a level of perseverance that he had rarely needed to muster in his past.

So why do some students succeed at musical training, despite the amount of effort and time needed, while others give up? Cognitive ability (the ability to reason, plan, problem-solve, engage in abstract thinking, comprehend complex ideas, learn from experience, and so on) is certainly important in achieving such pursuits. However, work by Angela Duckworth, a professor of psychology at the University of Pennsylvania and a MacArthur "Genius Grant" fellow, and her colleagues examined the possibility that people who have a combination of passion and perseverance, a noncognitive trait that the study referred to as "grit," tend to succeed at long-term goals better than those who don't.[3] The group studied large numbers of subjects in several different contexts: levels of educational attainment among two groups of adults, grade point averages of Ivy League undergraduate students, class retention of cadets at the West Point United States Military Academy, and rankings in the United States

National Spelling Bee. They found that grit had an incremental but potentially more significant role in predicting success than IQ (the "intelligence quotient," which roughly relates to some aspects of cognitive ability) and also than another trait, conscientiousness (the desire to do a task well and to take obligations to others seriously). Thus, even though grit may play only a small role in determining one's success at achieving a long-term goal like learning to play the piano as an adult, it may be a trait that pushes highly motivated, conscientious individuals with excellent cognitive ability to make progress toward their desired achievements.

Can a trait like grit be learned? Dr. Duckworth's group examined some of the factors that determine one's level of grit and reported that individual differences in grit may derive in part from differences in what makes people happy.[4] This makes sense, given that grit depends on passion, and we tend to derive happiness from engaging in activities in which we have a passionate level of interest. Genetics, however, may also contribute to grit as a trait. Rimfeld and colleagues studied 2,321 pairs of teenage twins (to account for nearly identical genetic factors) to try to understand the origins of grit and other traits.[5] The investigators examined the academic performance of these 4,642 teens and found that grit, like other personality traits, appears to be at least partly inherited and, as suggested by Dr. Duckworth's studies, does not predict success to a greater degree than other traits like conscientiousness. Another study by Tucker-Drob and colleagues examining twins in Texas came to a similar conclusion.[6] Nonetheless, given that grit is a combination of passion and perseverance, people who learn perseverance early in life and also have a genetic or environmental disposition that helps drive their passion for something may be better equipped later in life for achieving difficult long-term goals.

MUSIC PRACTICE IS CURIOUSER AND CURIOUSER!

Another quality of successful music learning is curiosity. Just as it promotes the motivation to compose new music as discussed in the previous movement, curiosity can also motivate how we learn to play or sing, and how we practice. Curiosity can drive music learners to ask questions such as:

- What happens when I move my fingers a certain way?
- Which positions are best for my hands?
- How is one part of the music connected to another?
- How can I make this piece of music easier to play or sing?

Numerous studies support the idea that curiosity is associated with better learning outcomes,[7] although the degree to which curiosity boosts learning may vary and may depend on the context.[8] While the brain circuitry behind how curiosity contributes to learning is unclear, several studies support the idea that curiosity enhances learning by activating areas of the brain involved with memory. For example, in a study where subjects were presented with trivia questions, the parahippocampal gyrus and the inferior frontal gyrus, areas related to long-term memory consolidation, became activated when a subject answered a question incorrectly but was then told the right answer. The subjects became curious about the correct answer and then learned the answer.

Dennis's successful sixty-year-old student had the passion for the music and had learned perseverance through his athletic training and teaching. He was also curious about how music works, which motivated him to learn. Dennis provided him with the strategy needed to maintain his curiosity and the motivation to improve. The student who walked out, although initially curious about the experience of learning music, had not previously found his goals exceptionally challenging and therefore had more limited experience with mustering substantial perseverance.

TEACH YOUR MOTIVATED CHILDREN WELL

As mentioned, from decades of teaching students of all ages, Dennis has learned that adults (himself included when he wants to learn a different instrument) have little patience. Adults seem to develop groundless expectations and high needs of instant gratification, forgetting how high-quality activities take devoted, focused time. When we are younger we are less time bound and accept ourselves more easily. A motivated eight-year-old and even a fourteen-year-old are more about wanting the experience of playing and developing skills for music making—so long as the motivation is their own and not the result of their parent's long-repressed desire to be a concert flugelhorn player. Not that there's anything wrong with flugelhorn players.

Dennis has noticed other differences in children's and adults' learning styles. For example, though adults take lessons in order to play, when they arrive for lessons they invariably sit down and start talking, explaining their week of practice and how it went and what the teacher can expect to hear when they do begin playing. Dennis never asks for this information in his lessons. He wants to hear them play (and not, in some cases, drone on and on about every detail of their week with their instrument as the lesson time ebbs away). A music teacher's role is activated by responding to the music itself, much more than to words about music. A teacher can assess performance with helpful and encouraging words while ascertaining what's needed for the following week's assignment. In contrast, children almost always come to a lesson so eager to play they can hardly wait to demonstrate their musical accomplishments. They have little time to slow down to talk about process or theory. Dennis generally finds that children rarely do the practice steps. Instead, they try to play pieces by repeatedly slamming through them, almost never willing to practice slowly.

For children, this has little to do with patience or thinking of time spent but rather their excited and excitable way of going about learning. So with this population, Dennis's strategy involves showing them "puzzle pieces" of the practicing process. He helps them understand that practice involves slowly and carefully connecting the pieces of music that are being learned, toward a successful, complete puzzle. His goal for these younger students, regardless of age, is to ask throughout a practice period, "What are you trying to accomplish right now?" Students are expected to give specific answers. Dennis loves responses like these:

- "I'm working on measures such and such so I can master the rhythm."
- "I'm working on hearing the thrust of this particular phrase so it has a climax."
- "I'm going to be conscious of the dynamics as I play this part."
- "I'm studying the form of the piece and its organization."

Of course, most children do not speak this way (and it would be a little scary if they did). In the early stages of learning the answer to this question will more likely be "I don't know" or "I want to play better." Nonetheless, coming to understand the reason for assembling these puzzle pieces is effective practice at any age.

Coaches apply similar ideas when working with athletes. After years of coaching youth baseball and softball, for example, Larry learned that to play against another team, a coach doesn't just send athletes forth and say, "Play ball." A coach develops strategies to teach separate aspects of the game (in the case of baseball and softball: hitting, throwing, catching, fielding), plus warmups (which in music can be scales and technique exercises). Finishing a lesson with a favorite piece of music that the learner can play in its entirety is most rewarding, and equivalent to playing that first complete game against another team. The process, if effective, always contributes to the desired outcome. Studying a musical instrument, therefore, can be a valuable process beyond the pleasure of the skill itself because it tends to not only teach us an expressive skill but to also develop the traits of successful lifelong learners.

UNEXPECTED BENEFITS: THE TRANSFERABILITY OF MUSIC LEARNING

In a separate survey of people learning a musical instrument (appendix B), we asked how they learned, what processes they experienced during music lessons, and how they felt about the lessons. Also, what motivated them to take lessons in the first place, and what were their relationships with their teachers like? What had they discovered about themselves through lessons and practicing? What had they learned about life through musical involvement?

Many reported a variety of unique benefits from their music lessons and learning how to practice. Here are some samples:

- "There is something rewarding about producing the sound that you desire."
- "Hearing the teacher say: 'Imagine you're kneading dough while playing this passage' makes a lasting impression, for it reminds me of how attitudes shape music, life, and decision-making."
- "Preparing for performance is undoubtedly the final goal of lessons, yet being able to play a single line of notes on the piano to help accompany children singing caused me to feel I contributed in a special way."

- "The fun of figuring it out on my own with the help of a small book of easy tunes and a fingering chart provided a special feeling of accomplishment, tinged with surprise that it worked."
- "It was deeply satisfying to have played something really well, not just play the right notes."
- "The lessons changed me as a human being."
- "I learned to find patterns in what I am doing and to think more systematically."
- "I remember leaving every lesson, brimming with understanding and a thirst for knowledge, often excited to get home and practice that one piece or section to make it better. To this day I have a love for learning stemming from the music lessons."

According to another person reflecting on past and present feelings of musical involvement:

> There are many different reasons to play music. Some days it was for pure mastery of a difficult skill . . . the satisfaction you get from executing a fugue with precision. It's the same feeling of mastering any other skill . . . a martial art, sketching, baking. Another day it was a need to express—improvise . . . create . . . communicate. It was another language, one that lacked some of the frustrating constraints of words. Through this, I discovered pieces of myself. I discovered what I thought mattered, what I thought was cool, all through the abstract aesthetic of sound.

In each of these responses and others, music learners are expressing both the joy of accomplishment in their musical endeavors and also the realization that achieving any short-term goal in their musical training may lead to additional achievements in music and other pursuits. They share a clear response to their learning: they all *like* to learn, and they all *want* to learn more. This pattern of liking and wanting is an underlying mechanism of how our brains process pleasurable experiences.[9] You eat a piece of chocolate, and find that you like it. Because you like it, seeing chocolate (or even thinking about it) makes you want more chocolate. When you get more chocolate, you like it again. Music listening appears to work in the same way. You hear a song and you like it, leading you to want to hear it again. We will discuss the details and neurochemistry behind this process in the eighth movement.

It is unclear whether this liking-and-wanting cycle applies to music learning, but it is likely that when we achieve those short-term musical goals (such as learning to play a simple song on the piano with both hands), our brains undergo all the processes linked to liking what we experience, leading the brain to activate all the signals involved with wanting more. The reward of achievement promotes the motivation to keep going. Furthermore, it is exciting to speculate that these rewards in music motivate learners to attempt other pursuits they may have previously thought to be beyond them, leading to a whole new cycle of liking and wanting that motivates them to learn and thrive, as suggested by the respondents of our survey.

THE IMPORTANCE OF THE TEACHER-LEARNER RELATIONSHIP: SOUR OR SWEET?

As we discussed, great teachers like Dennis (and we know Dennis is great because we didn't have to pay any of his students to say so) provide a strategy for students learning how to play a musical instrument that enhances chances of success when combined with passion-driven motivation and perseverance (grit), conscientiousness, and cognitive ability. However, a number of respondents to our survey recalled sad and scary experiences caused by poor teacher-student relationships.

Not all teachers understand human nature. Some music instructors are more focused on tasks than on the people who come to them to learn, and this focus is demonstrated by inflexible work styles and strategies. The attitudes of students toward practicing can reflect such approaches to teaching. One respondent stated, "I despise practicing. I seem to learn best when I'm by myself experimenting or problem-solving things my own way." Another complained, "I didn't love practicing—who does?" while another wrote, "Practicing was not a joy, necessarily; it was just something one did."

In contrast, for those fortunate enough to find a skilled, positive, encouraging teacher, learning to play an instrument will prove to be of incomparable value. For example, one respondent stated, "The teacher worked tirelessly with me on technique and I found that I loved working on technique! I learned that I had perseverance. I liked experiencing improvement in my ability and improvement in the sounds coming out of the instrument." Another shared: "I learned the importance of not starting the piece over when I made a mistake, but instead, going over and over the 'stuck' places until I did it correctly with consistency. It felt a bit like magic!"

Relational elements govern much of potential learning. One respondent stated this idea openly:

> The teachers I learned the most from were the teachers who were able to adapt my interests and incorporate them into the lesson plan, usually on the fly. How the teacher took interest in me as a person definitely changed my music trajectory. Teachers who asked lots of questions started me to process and think for myself. Then, practicing became enjoyable. I started learning how to listen and how to work on problems and come up with solutions. I began forming my own ideas about how to flesh out the notes on the page. Practice was no longer a necessary evil to be endured, but rather, my friend. Since being *motivated from within*, the practicing I do has yielded not only technical progress and expanded my repertoire, but it has taught me much about myself and the wonder of music. It has allowed me to share music with others in much deeper, more meaningful ways.

Others wrote, "I am surprised every time a new lesson works itself out in my practicing and I gain new competence. What I learn over and over is patience and humility." "Music was social and fun" and "I don't remember practicing with anything but pleasure."

It seems that multiple variables contribute to how music learners respond to their experiences with lessons and practice, but these variables are only one part of the formula. Is the teacher working to provide a flexible strategy to learn? Does the student have the passion for music that motivates their learning? Can the student use their skills in perseverance to celebrate small achievements? Understanding the answers to these questions could help refine strategies for teaching and learning that would better enhance the motivation to improve.

CONCENTRATION: ANOTHER KEY TO SUCCESSFUL MUSIC LEARNING

Concentration—the deliberate attempt to focus during tasks we perceive to be difficult—is another significant component of learning to play an instrument. As people concentrate, they engage more with a task with the purpose of maintaining a desirable level of performance. For example, when you ride a bicycle along open roads without traffic, you may pay greater attention to the music playing in your headphones, maybe even singing along with abandon. But in dense traffic, you pay greater attention to your surroundings and may not even realize a song is playing or may even choose to not listen to music at all. So when concentration is needed, our brains process fewer peripheral stimuli, to shield against distraction.[10]

Our respondents emphasized the merit of concentrating and trusting the process of mindful repetition. For example, one music learner stated, "I read—at sight—music I had never seen before. This facility is termed sight-reading. It takes maximum concentration and tenacity. I set the metronome (a constant beat) and stipulated that I would not stop, regardless. I played the same piece six times repetitively, mistakes and all, but kept my place in time. I was amazed at how I continually got better and felt like my brain was taking in much more than I was conscious of. Without stopping to practice the tougher spots they were ironing out themselves." Understand, this person may still "practice" tough spots, but this approach demonstrates one important form of music learning—out of countless approaches to musical execution and expression.

Another respondent, a psychologist, suggested that, as in the practice of psychology, music practice and performance had both conscious and unconscious aspects:

> When I think of the comparison between music and my field of psychology, two things come to mind: the conscious and unconscious. The conscious, the things we are aware of, and the unconscious, the things we are not aware of that still influence our decisions, our choices, and our actions. I think of how the notes on the page are similar to the words that I might say to a patient. The technicality of how to play music (like how I sit, how I move my fingers, how I move my wrists) is like the theory that grounds me to my theoretical conceptualization of a patient. But the unconscious feelings that drive how I interact with another human being are like the music behind the notes. The feeling of connection or disconnection to another being is the unconscious feeling we experience from music and the way it moves us. I could be saying the correct words to a patient in a moment of crisis, or

playing the right notes on the piano, but if the music behind the notes is missing, the willingness to emotionally connect the words I am saying to the patient is just noise. The same way notes without the music are just noise.

COLLABORATION: SHARED EMOTIONAL RESPONSES AND TRAITS IN MUSIC TEACHERS AND LEARNERS

Dennis's experience, the responses to our survey, and studies on the psychology and neuroscience behind learning to perform a difficult task—to pilot an airplane, ride a bicycle, speak a language, plan and grow a successful garden, or learn a musical instrument—show that success in learning requires motivation, determination, concentration, and usually guidance from another person. The person offering guidance must work with the student to find the best way to learn and improve. Learning to play a musical instrument in particular requires simultaneously coordinating many functions that are regulated by the brain, including:

- *motor responses* (such as fine and gross movements of your fingers, hands, arms, and even feet);
- *sensory input* (including, for example, information about the position of fingers or feet, the amount of pressure fingers or feet apply to an instrument, the visual input from reading a page of music, or auditory input to monitor playing);

- *emotional responses* (from joy when successful to frustrated when not, and also in response to the music itself);
- *analytical thinking* ("Am I playing that right?"); and
- *evaluative thinking* ("Hey! I sound pretty great!" or "I hope nobody heard that!").

If this sounds overwhelming, there is good reason: it is. Until you master each element on its own, successfully combining elements is nearly impossible. Enjoyment comes from the challenge and realization of how powerfully each grasped element can bring a final product to fruition. Mastering a piece can be an ongoing process of unfolding its meaning, in terms of both personal expression and the composer's intention. It's a search. It's never boring, but always challenging in positive ways, for music holds poetic space.

Fourth Movement

PRACTICING MUSIC, PART II

UNDERSTANDING THE NEUROSCIENCE

AS WE saw in the last movement, getting to the point where students can play a musical instrument well requires students to have certain traits combined with a learning strategy. And to get better, strategies must include practicing over long periods of time, whether in group rehearsals or working alone. Learning a piece of music and singing in a group for fun, not necessarily for a performance, are also forms of practice. At the other extreme is isolated practicing, which can be preparation for a solo event or seriously working on a part in a duet or for a larger group like an orchestra.

Practicing is more than playing a tune over and over. Effective practicing is exceptionally thought-provoking, involved work. Always asking "What am I, right now, trying to accomplish?" is a great way to start because it can help you identify and isolate the elements of a piece. Naturally, the need to practice and then to practice again and again is linked to the fact that, as we've seen, playing a musical instrument involves substantial demands on the human brain, including sensory, motor, cognitive, and emotional activities. Learning vocal music places similar demands on our brains. Indeed, it could be argued that learning to play a musical instrument is among the most difficult tasks a human brain can undertake because of the tremendous cognitive, fine and gross motor, and sensory demands that are required. Playing an instrument while singing takes those demands to a whole new level.

To fully understand the demands that music practice places on our brains, let us consider all the processes the human nervous system must undergo in someone learning to play, for example, the piano (though similar processes occur learning any instrument). This student, whom we'll call Portia (and who, like the so-named characters in Shakespeare's *Merchant of Venice* and Suzanne Collins's *The Hunger Games*, is creative and brilliant), has already had a year of lessons. She is learning a piece of written music, sitting diligently at the piano, hands at both starting positions on the keyboard, feet ready to use the appropriate pedals. She gazes upon the sheet music in front of her, which starts with an eighth note (middle C) that she must press with the middle finger of her right hand. Here are the steps Portia's nervous system has to go through to make that happen:

STEP 1: FROM PHOTONS TO NEURONS: TURNING ON A NEURAL CIRCUIT IN THE EYES WITH LIGHT

As Portia gazes at the written music on a page or screen, photons are reflected (off the page) or emitted (from the screen) and absorbed by the physical objects around Portia and Portia herself.

Photons are the basic units of light: always in motion and traveling at speeds as high as 2.998×10^8 meters per second (that's about 1,077,616,742 kilometers per hour or 669,600,000 miles per hour). In fact, right now, photons are moving in waves, reflected off the page you are reading or emitted from the screen you are gazing into, penetrating the outer portions of your eyes (including the cornea and lens) and a group of transparent neurons and other cells at the back of your eyes (in the retina), and finally stimulating specialized neurons in the back of the retina called photoreceptors (figure 4.1).

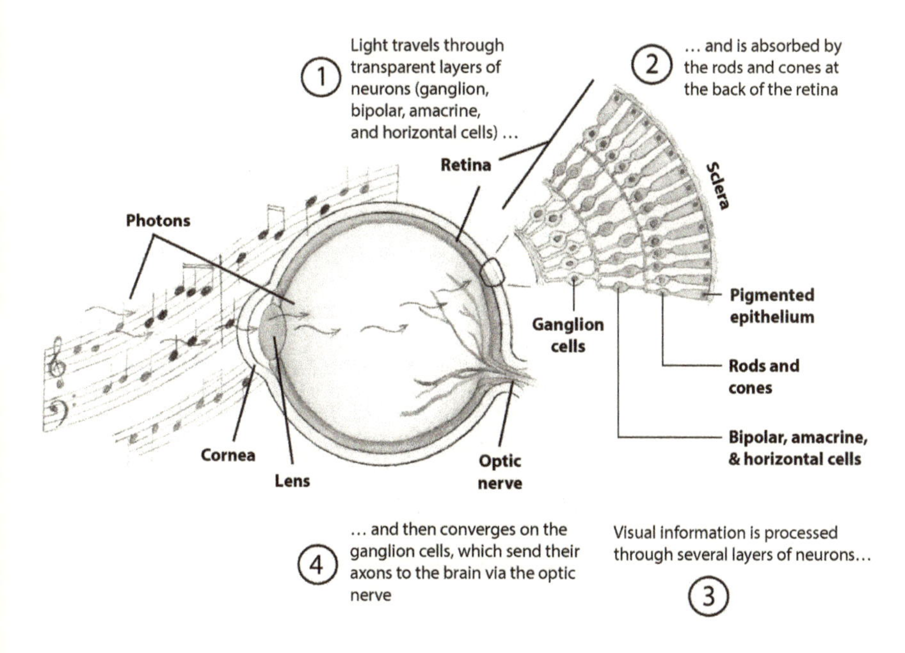

FIGURE 4.1 How photons reflect off sheet music and penetrate our eyes, stimulate photoreceptors (rods and cones) at the back of the retina, and then transmit their signals via bipolar cells and amacrine cells to ganglion cells, whose axons form the optic nerve. (Illustration by Susi Davis)

The human eye has two types of photoreceptors: *rods* and *cones*. Rods are responsible for vision at low light levels and are not involved in color perception. They do not help us detect fine details. Cones, on the other hand, are active at higher light levels, are capable of color vision, and are responsible for detecting all the little details that our eyes can see, like the flags on an eighth note (♪). So, like your cones, Portia's are activated by all the photons in front of her, including those coming from the written music she is attempting to read.

Once the cones are activated and fire their impulses, they transmit this information to those transparent neurons in the retina that we mentioned (*bipolar cells* and *amacrine cells*), that in turn activate specialized neurons called *ganglion cells* (figure 4.1). The connections between these cells are the first part of the circuit that deals with visual information.

STEP 2: GANGLION CELLS SEND SIGNALS TO OUR VISUAL CORTEX TO SORT OUT THE FEATURES OF VISUAL INFORMATION

The axons from all the ganglion cells (about 1.5 million of them in each of Portia's eyes) come together and exit the eyeball at a place in the retina called the optic disc. The optic disc is also called the blind spot since there are no photoreceptors in this part of the retina (there is no room for them with all those axons exiting). This huge collection of axons forms a cable called the *optic nerve*.

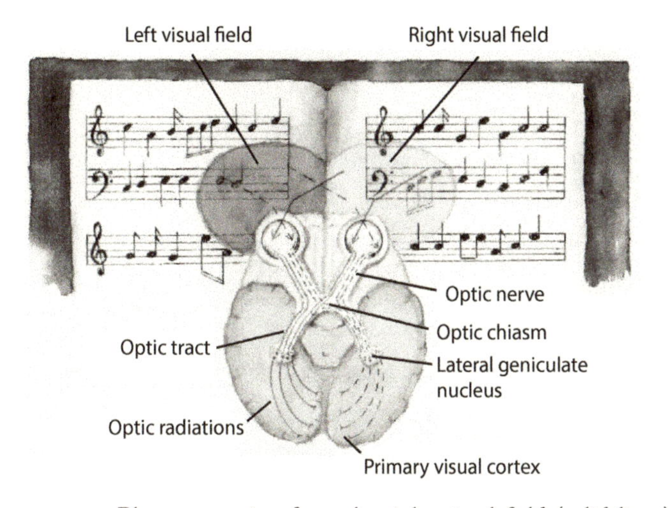

FIGURE 4.2 Photons coming from the right visual field (solid lines) and left visual field (dashed lines) are processed by both eyes. Photoreceptors and other cells stimulated by these photons in the retina send their axons to the lateral geniculate nucleus (LGN) in the thalamus. LGN neurons project their axons (collectively called optic radiations) to the visual cortex, where the brain sorts out the features of visual information. (Illustration by Susi Davis)

The optic nerve from each eye branches off, with some axons staying on the same side of the brain and others crossing over to join the opposite optic nerve at a crossing point called the *optic chiasm* (some also call it the optic chiasma; figure 4.2). This crossing allows each eye's visual information to be represented in both sides of the brain.

After the chiasm, these cables of axons go to different areas, including the thalamus (which, you may remember from the first movement, is located just above the brainstem and serves as a relay station for nearly all sensory information coming into the brain). In particular, ganglion cell axons transmit signals to neurons in a special part of the thalamus called the *lateral geniculate nucleus* (often referred to as LGN).

In the lateral geniculate nucleus, visual information is processed and then carried by lateral geniculate nucleus axons to the *primary visual cortex* at the back of your head (in the occipital lobe), which helps to sort out visual information. For example, different sets of neurons in the visual cortex detect different aspects of the visual world, including the shape, size, color, and location of an object and, if that object is moving, its speed and direction of movement.

STEP 3: OUR "HIGHER" BRAIN INTERPRETS THE SORTED VISUAL INFORMATION TO DETERMINE THE "WHAT" AND "WHERE"

Now that Portia's brain has sorted the visual information generated by those photons reflected or emitted from the written music in front of her, she needs to understand what it all means. To do this, her visual cortex cells send messages to other "higher" brain areas. In particular, some visual cortex neurons in Portia's brain send axons to the part where memories about an object's form and its name are processed (the temporal lobe).

Some people refer to this as the "what" pathway because it is critical for identifying objects—like the notes on a page of sheet music. Interestingly, the neurons involved in music reading and comprehension appear to be distinct from the neurons involved in reading words.[1] For Portia, the "what" is an eighth note. A second collection of axons from her visual cortex connect with the part of the brain thought to be involved with perceiving spatial relationships between objects as well as motion (the parietal lobe). This pathway is sometimes called the "where" pathway because of its involvement in analyzing how objects move and determining the location of objects relative to one another, like the position of notes on the treble clef and the bass clef. The "where" helps Portia figure out that the eighth note is middle C on the treble clef.

STEP 4: WE DECIDE TO PLAY

Once Portia's brain has decoded the symbols on her sheet music, its next task is figuring out what to do with that information.

How the brain makes such decisions has occupied the thoughts of philosophers and scientists for centuries. For example, seventeenth-century mathematician Blaise Pascal suggested that when we make decisions our brains evaluate the "expected value" of something by multiplying its value (that is, how much we perceive that we need or want it) with the probability that we can actually get it.

In this example, Portia must decide if the written music she sees in front of her is something she wants to play and whether she has the ability to play it. Although multiple areas of the brain are involved in making such decisions, two places in the frontal lobe (the orbitofrontal cortex and ventrolateral prefrontal cortex) appear to be essential for making decisions. In Portia's brain, these areas communicate with one another to make the assessments that Pascal predicted over 360 years ago.[2]

As a result of these areas interacting with one another and with other areas of Portia's brain, she could decide, "This looks hard; I'm not sure I'm ready to play this," which could lead to a different decision, such as turning to something else that offers an immediate reward, like going to her kitchen and getting a brownie. (Your brain, reading this, may similarly be processing thoughts loosely translated to "Mmmmm . . . brownies!") These types of decisions may moderate our diligence for learning music (or reading the rest of this movement) versus our tendency to procrastinate,[3] a behavior that often involves doing one thing that is easy or pleasurable (brownies) to delay doing something that is hard.

STEP 5: OUR BRAIN CONVERTS WRITTEN MUSIC INTO COMMANDS THAT CONTROL MOVEMENT

Since Portia has all the traits of a successful learner and has a great teacher (like Dennis) who has given her a strategy to learn to play this piece of music, she decides to skip the brownie (for now) and play.

So Portia's brain must translate that visual information into commands that regulate the positions of all her fingers on both hands (on or above specific keys on the keyboard), the amount of weight and velocity each finger will use to engage each key, and the amount of time each finger will hold the key down. At the same time, her brain needs to instruct her feet whether or not to depress a pedal, to position a foot over a specific pedal, how much pressure to place on the pedal, and how long to keep the pedal depressed. Her brain must also be prepared to change all of those parameters in preparation for the next set of commands dictated by the music on a page or screen.

All of these functions start in a part of the frontal lobe called the prefrontal cortex. Our "executive functions," like thinking about something, take place in this part of our brains. In Portia's brain, the prefrontal cortex receives the idea that the middle finger of the right hand should play middle C on the piano keyboard for a period that is an eighth as long as a whole note. The prefrontal cortex then communicates this idea to another part of the frontal lobe, the premotor cortex, and together they send signals to groups of neurons in the middle of the brain called the *basal ganglia*.

The basal ganglia make decisions about the appropriate form of voluntary movement (should Portia move her arm and finger to the left? should she stop moving her other fingers?) and then send signals to our old friend—the major relay center in the brain—the thalamus. The thalamus decides, based on the signals from the basal ganglia, whether to send an excitatory or inhibitory message to another part of the frontal lobe, the primary motor cortex, telling it what to do ("tell the muscles in the right arm, hand and fingers to play the middle C" or "don't play that middle C"). The axons of neurons from the motor cortex (motoneurons) then transmit signals from the motor cortex to the brainstem through the medulla (see figure 1.1 for reference). When these axons get to the medulla, they cross over to the other side of the brainstem and then go down the spinal cord, where they form synapses with other motoneurons that project axons through the arms and hands to muscles that result in a specific finger playing middle C with a specific amount of pressure for a specific duration (figure 4.3).

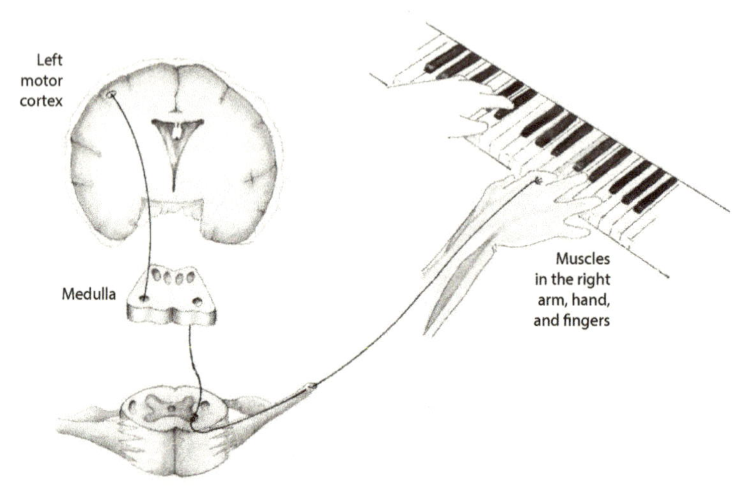

FIGURE 4.3 How motoneurons in the brain signal muscles in the right hand to play a note on the piano: Signals travel from (1) motoneurons in the left motor cortex (shown here as a "slice" through the brain) through (2) axons down to the brainstem and (3) cross over to the other side of the brain in the medulla, then descend through (4) the spinal cord, where they synapse with (5) neurons that send their axons from the spinal cord, down the right arm, and to the hand and fingers, signaling all of the movements to press a key on the keyboard. (Illustration by Susi Davis)

STEP 6: OUR BRAIN SENSES WHAT HAS BEEN PLAYED AND ADJUSTS TO IT

Portia's brain has now successfully converted a visual stimulus, originating from photons coming off music on a page or screen, to the action of playing a note on a piano. But we need to acknowledge the role of one more part of Portia's nervous system that allowed her to achieve this remarkable feat. While information is going from Portia's motor cortex to the muscles in her arms, hands, and fingers at a specific time just to depress middle C on the piano keyboard, information *from* the arms, hands, fingers, eyes, and ears must be transmitted to the brain both before, during, and after depressing middle C: first to let the brain know where Portia's fingers need to go relative to where they were before and then to tell the brain all the commands from the motor cortex have happened as planned. Her brain assesses that information based on what she senses with her fingertips, how her hands and arms are positioned, and what she hears.

The sensation of touching a piano key is transmitted from Portia's fingers through specialized structures under the skin at the ends of dendrites from unique neurons called sensory neurons. The endings of these sensory neurons are called *mechanoreceptors*, which means they send signals to the brain about mechanical changes detected at the surface of the skin.

Different types of mechanoreceptors detect different sensations, depending on the type of skin (for example, hairy forearms compared with the smooth skin on fingertips). Pacinian corpuscles on the fingertips provide the subtle sensation left by a small amount of pressure and by vibrations. On the other hand (well, the same hand and fingers in this case), Meissner corpuscles detect light touch and textures.

Upon touching middle C, here are the neurons in the circuit directing her reaction:

- *First neuron group:* Signals from Pacinian corpuscles and Meissner corpuscles in Portia's finger transmit signals along dendrites that travel up the hand and arm and into a little bundle of sensory neuron cell bodies just adjacent to the spinal cord called the *dorsal root ganglia* (figure 4.4).

- *Second neuron group:* Axons from these sensory neurons in the dorsal root ganglia enter a part of the spinal cord call the *dorsal column* (which is literally a column of axons from all the different sensory neurons) and then ascend up the spinal cord to the *medulla* and form a synapse with collections of neurons called the *dorsal column nuclei*. The neurons whose axons continue on from this point are the second group of neurons in this circuit.
- *Third neuron group:* The axons from the neurons in the dorsal column nuclei cross over to the other side of the brain via an area referred to as a *decussation* (which just means crossing) and then connect with neurons in a special part of the thalamus, the ventral posterior nucleus, which starts to process the sensory information coming from Portia's fingers. Neurons from the ventral posterior nucleus then make connections with neurons, the third group in this circuit, in the *somatosensory cortex* in the parietal lobe (figure 4.4).

To help us perceive what we are sensing, somatosensory cortex neurons communicate the feel of a piano key and how hard it is being depressed. The somatosensory cortex also receives inputs from another type of receptor, proprioceptors, which are sensory neuron endings in muscles, tendons, and joints that relay information about the position of all the parts of our bodies in space. Networks of cells in the somatosensory cortex interact with certain areas of the brain to help the motor cortex understand how muscles need to react to move Portia's finger to the position of middle C, to confirm that her finger is in the right place and plays the note with the right amount of pressure and for the right duration, and to help the motor cortex understand the location of that finger (and her hands and other fingers) in preparation for playing the next notes.

In addition to the mechanoreceptor and proprioceptor information being processed in Portia's somatosensory cortex, Portia's motor cortex can also be influenced by what she sees and what she hears. What's more, while interpreting a visual stimulus like written music, which we have already covered, Portia may also be looking at her hands from time to time while she is practicing. This visual information can be relayed to the motor cortex to help Portia adjust her positioning. As we discussed in the first movement, auditory information is processed through networks of neurons between the auditory cortex, the motor cortex, and other areas of the brain. These can also influence how Portia plays, helping her determine how quickly and how long to press a key.

All of this sensory information is also processed in another area of the brain, the cerebellum (figure 1.1). The cerebellum is linked to almost all of the other regions of the brain and contributes to learning as well as controlling the timing and fine-tuning of movements. Thus, Portia's cerebellum is instrumental in helping her learn how to move her hands and fingers to generate the sounds dictated by the photons coming off the written music perceived by her eyes.

So now you have an idea of what the brain must do when practicing to play a middle C on a piano (or any other note on

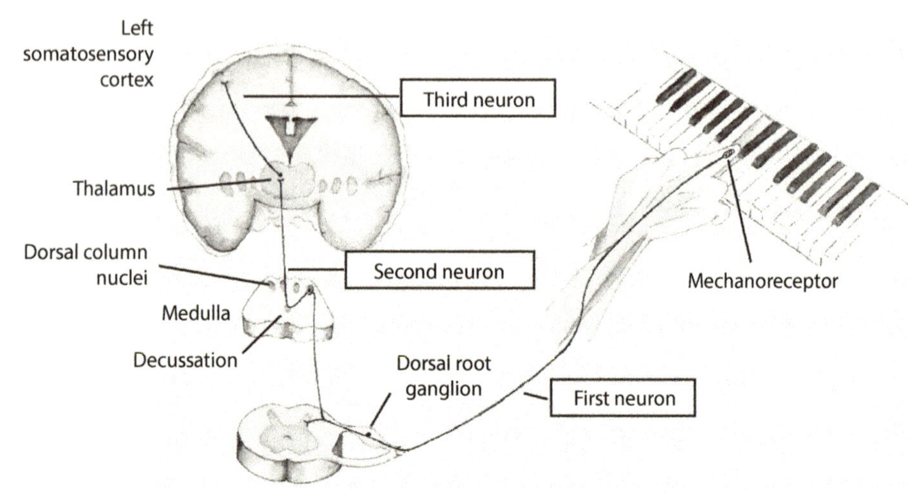

FIGURE 4.4 How the brain processes touching a piano key. Numerous mechanoreceptors in the fingertip send signals via the dendrites of sensory neurons (first neuron group) to the dorsal root ganglia, which travel through the spinal cord to the medulla and synapse in the circuit in the medulla (second neuron group). Axons from this second neuron cross over to the other side of the medulla and then synapse in the circuit in the thalamus (third neuron group). From there, these neurons synapse with neurons where information is processed about how the piano key feels and how much pressure the finger is applying (in the somatosensory cortex in the parietal lobe). (Illustration by Susi Davis)

any other instrument) with the middle finger of your right hand in response to written music. And it must do everything that we have summarized here (and more!) in less than a second. Of course, most music (except possibly something by a minimalist composer like Philip Glass) involves more than playing a single note. So, multiply everything you have just read by a factor of ten (for those of you with all ten fingers) and a little more for your feet in case you need to depress a pedal from time to time, and add many bars of music. Now you can appreciate why playing a musical instrument is one of the most challenging things a human brain can do.

BEST PRACTICES FOR MUSIC PRACTICING

Now that we have summarized the basic brain circuitry underlying music practice, we turn to the rich behavioral aspects of learning to perform. Given the remarkable challenge that music learning places on the brain, music students who want to maximize their musical expression must engage in formal, intentional practicing. This brings us to a special colloquialism—*woodshedding*, so named because a woodshed, often a distance from the main house, is a place to be alone for focused work. For some, practicing means a special place and perhaps a preferred time. For others, place and time are not important. But in any case, what's needed is a dedicated space where mistakes are a welcome part of the learning process, a safe place to mess around, experimenting to see what works. This is where a student finds a way to smoothly put everything together, learning to process information from the eyes all the way to the motor cortex, and then to the vocal cords, hands and fingers, or feet and back again.

In Dennis's experience, too many students, young and old, do not practice enough or effectively. Having heard and seen many students practicing over the years, he has observed the tendency of beginning a piece and moving through it until a mistake occurs, and then redoing that passage and going on, or worse yet, going back to the very beginning and retrying. This trial-and-error approach mistakenly assumes that by simply fixing the problem one makes the piece better. For Portia, or any student, to make preventing an error in the first place the main goal is a negative approach to learning and is not ideal because every mistake played is a mistake learned.

It's ultimately more effective to identify a challenging part of the music and focus on that. For example, if the appropriate physical gesture does not achieve the desired sound, then address technique (for example, fingering). For troubles with rhythm, singing or tapping out the rhythm helps instill it in the brain. If Portia either misreads notation on the music or mistranslates the notation to her instrument, be it her voice or any number of possible instruments, then she needs to get her notation straight. What she hears must match what she's reading. Careful work on selected areas of pieces and the desire to understand the music carry much more wisdom as a process.

Focusing on one or more of these skills—technique, rhythm, notes—and digging in toward mastery gets closer to describing authentic practicing. This practicing can be done with or without a teacher, although, as we discussed in the third movement, having a guide who works with music learners to devise practice steps and strategy can be helpful. In any case, a clear assignment and a clear desired outcome are key, assuming the learner actually practices instead of just playing through a piece over and over. Fantasizing that the piece is getting better through hurried repetition is the least efficient approach, yet the patience needed for slow and deliberate practicing is often in short supply.

As mentioned, when making music, one must be listening and simultaneously making necessary adjustments to the various music elements. Timbre, as an example, can govern accumulated hours of rehearsal time in the woodshed. As Philip Ball writes in his book *The Music Instinct*, "Timbre is arguably the most personal characteristic of music. When it comes to singing, timbre often holds the key to our preferences. Billie Holiday, Frank Sinatra, Nina Simone, Bob Dylan, Tom Waits, Robert Plant, David Bowie—all have vocal timbres that are instantly recognizable, and it is this, more than choice of song, that is the defining quality of their sound. The emotive power of timbre is tremendous."[4] While certain aspects of vocalists' voices are genetic, how they practice over years of time will move them toward their unique sound.

A sometimes-overlooked facet to tone quality is projection. When Dennis was studying trumpet with Tom Stevens, a high-level player with the Los Angeles Philharmonic, Tom asked Dennis how far he wanted his sound to project: 25 feet out? 100 feet out? 1,000? Dennis's first reaction was "I play the trumpet wanting a desired tone. I have never thought about projection." Tom's direction to consider projection in different spaces took Dennis to an entirely different level of tone production and changed his practice strategy. That was a life-changing moment for Dennis, not just for his music, but in many life situations, especially verbal communication.

Another consideration for practicing involves musicianship: putting all of these sensory, motor, emotional, and cognitive elements together. To combine the various components of music for a meaningful performance, each needs to be individually explored and practiced. It's much like the way you would approach a puzzle. Seeing the picture of the puzzle before searching for puzzle pieces to piece together is a much more effective way to complete a puzzle than just randomly trying different pieces. So the more the music learner explores and understands each musical component, the finer the final rendition will sound. Toward that end, while learning a piece of music, asking questions like these can help learners fine-tune their performance:

- Where is the music going?
- What is the context for each note in a phrase?
- What is the overall reason for the piece?
- What makes me want to learn this piece?

Finding the answers to these questions as part of the process of practice will lead to a fuller picture of the musical puzzle.

Internally hearing the music you want to make is the priceless advantage of great musicians. How could Beethoven, for example, imagine and compose music when he was nearly or totally deaf? The answer is "audiation," the process of mentally hearing the music and understanding where it is going. It's a matter of hearing the music in your head and sensing its meaning and significance without the need to physically hear it performed. As we mentioned earlier, this is much like an interior designer seeing a room and imagining, in great detail, a totally different look—the potential, all without seeing it actualized.

Audiation is much the same, and you can experience it right now by trying this experiment: Run through "Happy Birthday" in your head without singing it out loud.

You can practice developing and applying that skill when performing from notation, playing by ear, improvising, or composing. Many musicians can practice without their instrument by audiating their piece. Audiation is what allows conductors, who have exceptional audiation abilities, and players to hear where the music is going. People hear more mindfully with expectation and are thus more aware of surprises and detours. Instead of hearing and performing notes, you hear and perform ideas that are linked. It is having a constant aural image of a desired sound. Cherish audiation, for it transforms a mechanically decent pushing of buttons (or keys, or singing of notes) into a soulful delivery.

Speaking of soulful interpretation, how do you move in that direction, assuming it is your desire? Dennis recalls being in the audience when a performer finished a piece and two people behind Dennis murmured that the beautiful dynamics of getting louder (crescendo) and softer (decrescendo) cannot be taught. Later that evening Dennis, speaking with the performer's teacher, relayed what he had heard from the people behind him. The teacher informed him that what these audience members heard from the performer was taught, worked on a lot, and learned.

Dennis had a similar experience with an eighteen-year-old student learning Chopin's *Fantaisie-Impromptu*, which begins with a stimulating, virtuosic display that in the middle settles into a romantic, tender melody of extreme contrast. This young man, full of energy (and, it seemed, hormones), respected Dennis and was trying to shift from the exuberant mood to the gorgeous melody in this much slower middle section. Nothing seemed to work, until Dennis asked the student if he had a class or teacher in school that he really detested. The student immediately began to tear into one of his teachers, ranting on about his wasted time and resources on that course. He spoke with such fervent passion that Dennis asked him to hang on to the feeling while he played the slow part of the piece. As the student did so, Dennis felt his eyes tearing up, as the intensity of the student's soul poured into the beautiful melody. What a difference from just playing the notes correctly; "correct" does not necessarily communicate the composer's intention.

Whether from a teacher leading the process or a person alone figuring out how to make every note speak meaningfully to an audience, this is the kind of work that true practice entails. It's deep, probing work.

Finally, other than students making sure they regularly practice, one of the most difficult obstacles is the invasion of thoughts or ideas that block them from learning. Portia faced this problem when considering the possibility that the music in front of her was beyond her ability to play and that getting up and going to the kitchen for a brownie would be a better use of her time. Dennis himself faced this situation when studying with a particular piano teacher. Dennis thought that he physically could not play a piece of music by Mozart, believing that his fingers did not have the mobility and velocity needed for a particular section of the music. It turned out to be a rhythmic problem. The teacher pointed out how the fingers would cooperate if Dennis gave them the security of the rhythmic process. Dennis concentrated on the precision of the rhythm in the piece, and immediately the technique was there. This was exciting! This is the type of problem solving that goes into intelligent practice. It also, again, speaks to the value of working on defined areas of the musical puzzle instead of the whole puzzle all at once.

WHAT MUSICIANS SAY ABOUT PRACTICING

Here is a sampling of responses from our survey. Musician Chet Boddy wrote, "When I am trying to memorize a piece, I am visualizing the lyrics and the story in my head. I am developing a muscle memory for the notes and chords.[5] I am learning to hear the entire piece in my head so I can play it back at any time, as if my brain were a kind of CD player. I often hear music in my head much better than I am able to play it." This quote illustrates how audiation is critical for musical practice.

Others describe how their brains drive the process of practicing music. As pianist Sylvia Gray put it, "I believe I am building connections and patterns, and that many parts of my brain are active—the rational, the emotive, and other things I know not what. It just feels disciplined in a good way and healthy and orderly." Vocalist Pamela Plimpton wrote, "My brain is practicing mindfulness! I am focusing, focusing, focusing. I also think that there is a sense of pleasure derived from this. Ultimately, practicing really hard, focusing for as long as I am being productive, pays off. Often the payoff is not until the next day, after

I've 'slept on it.' Whatever I've worked on has had a chance to settle in, to 'jell' in my brain and in my muscle memory as well." This response is interesting in light of over a century of research indicating that sleep is required for memory retention, as well as more recent evidence that sleep involves an active process of memory consolidation.[6]

Pop and jazz singer Valerie Day described her process of practicing in some detail:

> I begin by warming up the instrument (my voice or hands) with exercises to get the muscles loosened up and increase the circulation in them. I then sing or play through a piece of music that I'm working on. When I make a mistake, or don't achieve the sound quality I'd like in a passage—I go back and chunk out the section that's not working. Oftentimes I'll slow it down to give my brain and muscles time to learn the finer articulations, then will speed up the tempo to a place where I can barely sing or play with any precision at all. When I go back to the original tempo I'll have more control because my muscles have learned where to be and when.
>
> After going at an increased tempo it's easier when it's slowed down again.
>
> After I have a mental map of where I'm going and have created enough muscle memory to get there, I go back and sing the song with emotional intention to make sure I'm not singing technique, but actually communicating the song. Sometimes I'll try communicating it from different emotional places to see what changes (phrasing, tempo, timbre, articulation, intensity, volume, etc.) occur. When practicing,

> the goal for me is to create a solid container that the music can then inhabit. Failure is essential to practice in that it gives you feedback about where you are in relation to the goal you've set for yourself.

Valerie's statement again highlights the importance of memory, focus, experimentation, and the need for that woodshed where it is safe to make mistakes.

Fifth Movement

PRACTICING MUSIC, PART III

CHANGING YOUR BRAIN TO GET IT RIGHT

IN OUR survey, we asked musicians, composers, and music lovers what they felt was happening in their brains when they practice. The majority of people who answered believed serious changes were afoot. Katie Bowman, a tenor sax player and vocalist, wrote, "When I practice something I know I am creating, altering, or reinforcing neural pathways." Rick Olsen, a musician and music teacher, said that when he practices, there is a "fireworks show going on. Synapses in the left and right brain; senses fully at work." Musician and composer John Smith wrote, "There are various components: hand-eye coordination, hearing, the constant passage of time underneath the music," which, John felt, "the brain squishes all into a coherent whole." And jazz vocalist Thea Enos wrote that practice is "like creating a new road" in her brain: "first it is wilderness, then rough (dirt, gravel, potholes), then smoother, then eventually a sturdy highway."

These responses support the link between the act of practicing music and intense activity and changes in our brains. Our perceptions of brain alterations arise from the changes we see in ourselves following multiple sessions of practicing. If you are learning to play an instrument and have been effectively working on a piece of music for some time, you will see changes in motor, sensory, and cognitive functions as you continue to practice and take on new challenges.

PRACTICE MAKES EVERYTHING BETTER

Dennis has noted common changes occurring in music learners as they study music at an instrument. As we outlined in the fourth movement in the case of Portia, teaching piano requires pianists to read two lines of music and perform different actions with each hand. Even the relatively simple task of playing the same pitches an octave apart with both hands in unison presents a challenge because the hands align oppositely on the keyboard, with our thumbs on opposite ends. Such a seemingly simple task takes time to achieve with competence.

Let's consider a few other examples that Dennis has observed over the years. As we discussed in the fourth movement, three skills are essential areas of focus when practicing: technique, rhythm, and the ability to read and understand musical notation.

Imagine you are taking lessons from Dennis and he asks you to play a scale to the pulse of a metronome or your foot tapping, two notes per pulse. After a week, you return and play metronomically, virtually announcing each beat with an accent instead of smoothly playing the notes. At this stage, you are following rules but not really making music. Dennis then asks you to keep the metronomic accuracy without announcing your consciousness of the pulse. You proceed with strict attention to the pulse, but now play the scale flowingly and musically, not mechanically. The desire to be rhythmically accurate while also playing smoothly changed because of your new goal and purpose, driving the ability to combine rhythm with the fine motor movements needed to play the scale. All you needed was guidance and the time to develop those combined skills. Purposeful practicing requires that you communicate a message with "soul." In other words, it must include the correct notes and rhythm, but also feeling.

Fingering—the sequence of movements by the fingers used for a musical passage—provides another illustration of how we physically change when we practice. Similar to typing on a computer keyboard (or an ancient device that Larry and Dennis used in their younger days: the typewriter) where "hunting and pecking" is slow and inefficient, playing a piano with the wrong fingering prevents a smooth performance of the music.

Fortunately, sheet music often has marks that show which fingers to use, and by noticing and adhering to the suggested fingering, you can produce the composer's desired sound. Chopin's markings, for example, guide the pianist to make a connected line of sound. A musician might ignore the written fingering to more quickly render the written music onto the keyboard, but musically speaking it doesn't "sound right." Integrating Chopin's concept and the fingering to pull off a musical phrase as a phrase helps the musician understand and, eventually, reproduce the composer's intent. Students will start with the choppiness and with time, make the appropriate physical changes in their fingering to achieve the Chopin-ness (sorry!).

Given that students and even accomplished musicians demonstrate such clear changes in their ability to read music over time and translate what they read into performed music through advancements in motor, sensory, and cognitive abilities, we might expect that these changes are linked to profound structural changes in the brain along the lines of the road building that Thea Enos suggested. Three processes that influence the function and structure of the brain may be affected by the challenge of music practice: generating new neurons (neurogenesis); increasing the connections between newly generated and already existing neurons (synaptogenesis); and generating the myelin sheaths that surround axons and dramatically increase the speed of neural impulses (myelination) (figure 5.1).

PRACTICE MAKES . . . NEURONS?

While we have already discussed the circuits needed to convert the lines of music on a piece of paper to the movements needed in each hand to play that music, the question remains: What changes must occur in the brain to become proficient in reading and playing music? And are the types of changes Dennis has seen in students' performances over time due to physical changes in the nervous system?

If our brains must build new circuits as we practice to become proficient musicians, you might expect that we need to make

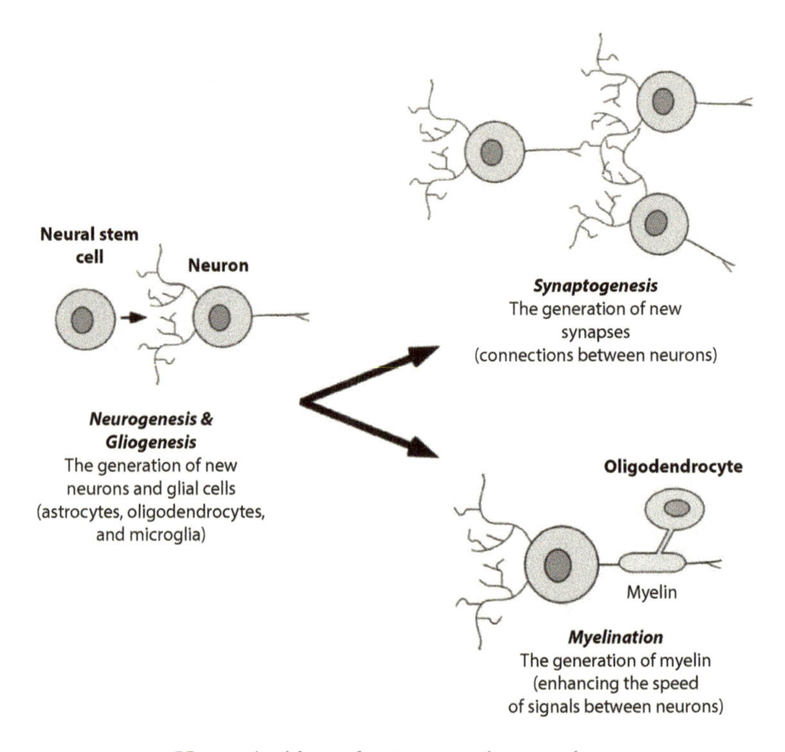

FIGURE 5.1 How to build—and maintain—the central nervous system. New cells (including neurons, astrocytes, and oligodendrocytes) arise from neural stem cells or other progenitor cells (through neurogenesis and gliogenesis). New neurons must connect with other cells by forming new synapses (synaptogenesis). Oligodendrocytes reach out to and wrap around the axons of many of these neurons and form myelin (myelination) to increase impulse speeds.

new neurons for these circuits. Back in the Dark Ages, when Larry was in graduate school, the general consensus among many neuroscientists about the possibility that our brains could make new neurons throughout life was that it just didn't happen. Unlike most cells, they reasoned, once a neuron is born its dividing days are over, except, possibly, under certain abnormal conditions (such as brain tumors). The only division a neuron experiences in its lifetime occurs when the cells that become neurons (neural "stem" or "progenitor" cells) divide. When neural stem cells become neurons, the process is called neurogenesis. Neural stem cells can also become other cells in the brain (collectively called glial cells—discussed later) through the process of gliogenesis (figure 5.1).

Neural stem cells are like young children in elementary school, capable of going on to do so many different things. As they go through school and eventually enter college or have other life-learning experiences, they start to differentiate in response to instructive signals that they are taught and other environmental influences. In most cases, they eventually turn into people with specific interests and skills and become professionals at some task. Neural stem cells similarly receive signals that drive them to become neurons or glial cells (figure 5.1).

It was believed that the dividing neural stem cells that become nondividing neurons only existed during embryonic and possibly early postnatal life. This idea was proposed in 1928 by Santiago Ramón y Cajal, who is often thought of as the father of modern neuroscience. However, thirty-four years later, Cajal's idea was challenged by the findings of Joseph Altman at the Massachusetts Institute of Technology. Using a substance that is only taken up by dividing cells, Altman found evidence of neurons dividing in the brains of rats with a minor brain injury.[1] This study, and several more by his group, led Dr. Altman to propose that dividing neural stem cells had taken up the marker of cell division and then differentiated into neurons.

In other words, Dr. Altman was proposing that neurogenesis—the generation of new neurons—didn't just occur in the developing brain, but also in the adult brain.

Dr. Altman's findings were exciting—some would say revolutionary!—and nearly nobody believed them.[2] In fact, his laboratory lost grant support because the neuroscience research community would not accept this heretical challenge to the Ramón y Cajal dogma. However, through the work of dozens of pioneering investigators who continued to pursue this research in many different species, the idea of neurogenesis in the adult brain was finally accepted in the 1990s.

In younger humans, neurogenesis can be detected in a few areas of the brain such as the hippocampus, a brain structure in the temporal lobe that we briefly mentioned in the second movement, so named because it looks like a *hippocampus*, a creature from Greek mythology with the head of a horse and the body of a fish (figure 5.2). It also looks like an actual seahorse, and the scientific name (genus) for seahorses is *Hippocampus*.

The hippocampus is important for learning and memory and for the associations between memories (like linking the beginning of spring to certain flowers blooming). Laboratory studies have demonstrated that neurogenesis in the hippocampus is necessary for certain types of learning and memory to occur, including learning motor behaviors.[3] As we push ourselves to learn and remember new types of movement (like playing scales with a certain rhythm), new neurons are born in the hippocampus that influence the activities of existing neural circuits or generate new neural circuits.

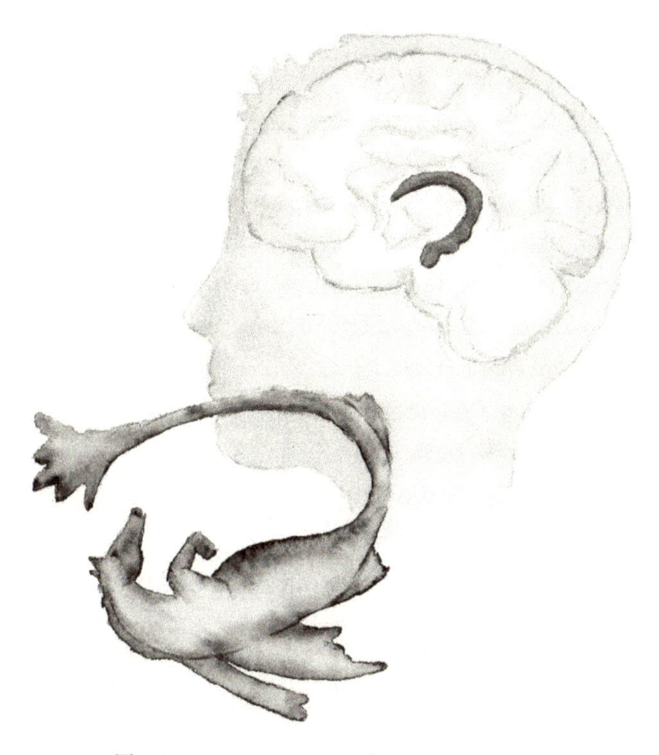

FIGURE 5.2 The hippocampus, shaped like the mythical creature, is located within the temporal lobe. (Illustration by Susi Davis)

As we age, neurogenesis declines to barely detectable levels. Nonetheless, even adding a very small number of new cells to the brain can dramatically impact the function and activity of any given neural circuit.[4] For example, if you have a bunch of ordinary lamps all plugged into the same outlet in your living room, turning on the switch just gives you ordinary, boring light. But now, if you plug into that same outlet a single extra cord connected to a disco ball, turning on that switch turns your living room into a party every time. The same is true in the human brain. Small changes in circuits can dramatically affect our behavior.

Although the hippocampus is undoubtedly important for learning music, there are no studies to date directly showing that practice promotes neurogenesis in the hippocampus or elsewhere in the brain. However, a number of studies have noted structural changes in the hippocampus consistent with more neurogenesis in musicians who engage in extensive practice. For example, one study found that parts of the hippocampi of expert pianists had a greater volume of gray matter (where neuron cell bodies are found) than the hippocampi of nonmusicians.[5] Similarly, another study of amateur musicians who practiced with a variety of instruments reported increased hippocampal volumes that increased with increasing musical experience.[6] These findings suggest that the more you practice, the more you grow parts of your hippocampus (within limits). One possible explanation for this increase is that practice induces neurogenesis.

What are the benefits of these changes in the hippocampus? A study by Oechslin and colleagues[7] suggested that increased hippocampal volumes in musicians (both professional and amateur) correlated with improved problem-solving skills and learning capabilities, although the findings from this study need to be confirmed.

Given the fact that cognitive, motor, and sensory systems are being thoroughly challenged in the process of practicing music (especially instrumental music), we speculate that neurogenesis likely plays a role in learning to play an instrument.

Furthermore, given that neurogenesis declines as we age, the changes in the rate of neurogenesis between young and old music students could at least partly explain the fact that older students often need to work harder to achieve some milestones in music learning compared to younger students. So perhaps we need to rephrase our statement about old dogs from the third movement: It's not that you can't teach the hippocampus of an old dog (or music student) new tricks; it's just that you need to do more to stimulate neurogenesis in the dog's (or music student's) brain.

PRACTICE MAKES CIRCUITS

One of the most remarkable discoveries in the past fifty years is the degree to which the human brain can undergo structural changes throughout life. When we learn new skills, like playing a musical instrument, that learning and the memories that are formed require the formation of new neural networks, such as the networks we discussed in the second movement, forming high-speed connections between many different brain areas.[8]

These networks develop as axons and dendrites from neurons both near and far from one another (either already in existence or generated through neurogenesis) create synapses that form new circuits or alter existing circuits. This process is called *synaptogenesis* (figure 5.1). A single neuron can have thousands of synaptic inputs coming from thousands of other neurons, and that single neuron can have thousands of synaptic outputs, influencing many different processes in the brain.

Some changes in existing synapses can occur over seconds or minutes, while the growth of new synapses may require hours or days. These changes reflect the remarkable level of plasticity in the human brain. When we learn to play a musical instrument, the memories for what we are learning are encoded through synapse formation in places that include the hippocampus. Almost immediately, synapses begin to form, leading to new circuits whose signals strengthen as we practice or rehearse a piece of music.

In the first movement, we discussed how many synapses are chemical in nature, passing chemical signals called neurotransmitters from the axon of one neuron to the dendrite of another in a neural circuit. The synapses that form as we are learning to play music are considered "excitatory" and utilize a particular neurotransmitter called *glutamate* (figure 5.3). An important concept underlying the cellular mechanisms for successful musical practice is that "neurons that fire together, wire together." As we practice, we increase the strength of excitatory synapses as they release glutamate into the space between axons and dendrites, leading other neurons in that new circuit to fire as well. In contrast, when we stop practicing or rehearsing, those synapses weaken and may eventually disappear.

Recall from the second movement that signaling in some neural networks strengthens while other networks are diminished in the brains of improvising jazz musicians. The same thing likely happens as we practice music. The excitatory synapses that form new circuits as we practice music can lead to other excitatory networks becoming strengthened but also to some being weakened to help focus our learning. Consider a music student who, before taking up guitar, spent considerable time pole vaulting. As this student becomes more engaged with music, spending many hours learning the guitar and fewer if any hours pole vaulting, the circuits required for pole vaulting will weaken as the guitar circuits strengthen (unless some of those circuits are required for both activities). In addition, some circuits may need to be actively turned down while learning the guitar, which could be achieved by inhibitory synapses that use inhibitory neurotransmitters, like gamma-aminobutyric acid (GABA) (figure 5.3). Evidence increasingly suggests that balanced combinations of synapses that use glutamate (excitatory) and GABA (inhibitory) are required for the neural plasticity underlying learning and memory.[9]

A number of studies support the idea that the brains of musicians who engage in extensive music practice are undergoing

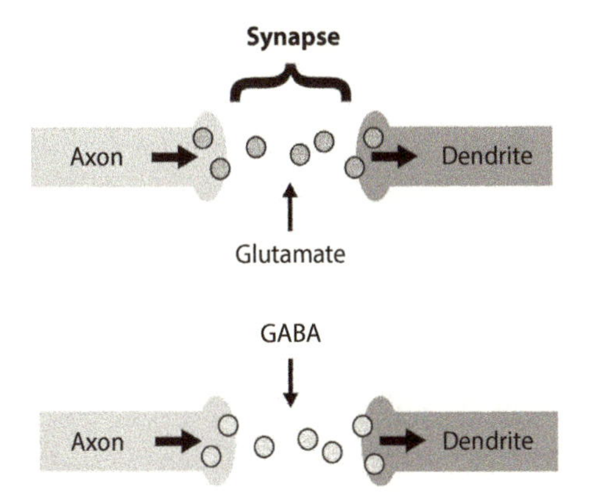

FIGURE 5.3 How neurons can release excitatory neurotransmitters (such as glutamate) or inhibitory neurotransmitters (such as GABA). Different combinations of excitatory and inhibitory cells in a circuit can regulate how that circuit functions.

more synaptogenesis—the creation of new synapses—compared with nonmusicians. Rosenkranz and colleagues, for example, used a procedure called transcranial magnetic stimulation (TMS) to measure motor-evoked potentials (MEPs).[10] If you are wondering what on earth that means, TMS involves delivering magnetic pulses to the brain via a coil that is held up to your scalp. An electric current is delivered to the coil, which induces an electrical current in the part of the brain nearest the coil. Motor-evoked potentials are the electrical responses of cells in response to the activity generated by motor neurons. So, if you use TMS to stimulate the motor neurons in the brain that lead to responses in a specific muscle in a finger, you can use MEPs as a measure of the synaptic connectivity in that particular circuit.

In the study by Rosenkranz and colleagues, the researchers used TMS to stimulate the activity of motor neurons that activate muscles in the fingers of female nonmusicians and musicians who were twenty to thirty-five years old. Most of the musicians were pianists who had all started playing at an early age, and they also included a guitarist, a trombone player, a trumpet player, and a recorder player. The results indicated much greater MEPs in musicians compared with nonmusicians and also indicated that the earlier the musicians had started practicing in life the greater the effect. These data are consistent with the idea that the more you practice music, the more synapses you create or strengthen.

Practicing music is therefore an effective way to promote plasticity in your brain. Whether it's learning to play that Chopin piece, a Carnatic composition by Purandara Dasa, a ballad by Roy Orbison, a piece of chamber music by Cacilda Borges Barbosa, or a great hip-hop song by Jay-Z, learning to practice and play music leads to the generation of new circuits that enhance your motor, sensory, and cognitive function. As we will discuss, these changes may not only take you from choppy to Chopin, but they may also improve how you move, sense, and think in general.

PRACTICE MAKES MYELIN

Neuroscientists spend most of their time using their brains to think about brains, including their own brains. Larry is no exception. In fact, he uses the networks of neurons in his brain to think about a substance called *myelin*. One of the most remarkable characteristics of the human brain is the speed at which different parts of the brain communicate with one another. This speed is especially important when we listen to, compose, or perform music. Although many neurons need to fire fast, others are required to fire at slower speeds. The difference between fast- and slow-firing neurons turns out to be myelin, a substance generated through the process of *myelination* (figure 5.1).

On average, nerve impulses along axons that lack myelin travel at about 1 meter per second (just over 2 miles per hour). In other words, impulses moving along such "naked" axons in your brain travel about as fast as a car on a freeway in Los Angeles at rush hour. When axons are enwrapped in myelin, their nerve impulses speed up to allow for the extremely fast communication needed between different areas of the brain for higher-order activities like engaging with music. Axons that have myelin conduct impulses about 200 times faster than "naked" axons. In other words, going from an unmyelinated axon to a myelinated axon is comparable to the difference between driving on a Los Angeles freeway at rush hour and driving on the autobahn in Germany at midnight with no speed limit or traffic.

Myelin is a beautiful structure that wraps around axons multiple times. Myelin owes its ability to increase the speed of nerve impulses along axons to structures called nodes of Ranvier (named after Dr. Louis-Antoine Ranvier, 1835–1922). These nodes are areas along axons where myelin is absent. Signals jump from one node to the next, dramatically increasing their speed (figure 5.4).

To illustrate how important myelin is for vertebrate animals like us, evidence suggests that the first animals to have myelin

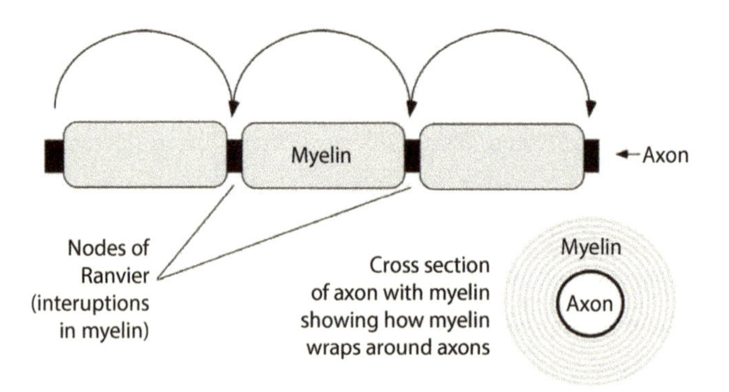

FIGURE 5.4 Myelin increases neuronal impulse speed. It wraps around axons many times in segments interrupted by nodes of Ranvier, allowing signals to jump (*curved arrows*) from node to node.

included a group of fish with hinged jaws called placoderms that swam around the earth about 416 to 359 million years ago.[11] With the advent of myelin, fish and the animals that followed were able to establish high-speed connections between different areas of their nervous systems, making them better predators and giving them the ability to respond to their world much more quickly than their predecessors.

In humans, myelin enables our brains to perform high-speed processes that allow us to quickly coordinate our movements (like playing a violin), perceive our environments with all of our senses (like listening to a concert), think, and, of course, enjoy and engage in all the different parts of music almost instantly when we are listening to or performing music.

In the brain and spinal cord, myelin is made by a type of glial cell called an *oligodendrocyte* (figure 5.5), which has many branches that wrap around axons several times. One oligodendrocyte can supply myelin for multiple axons. Although

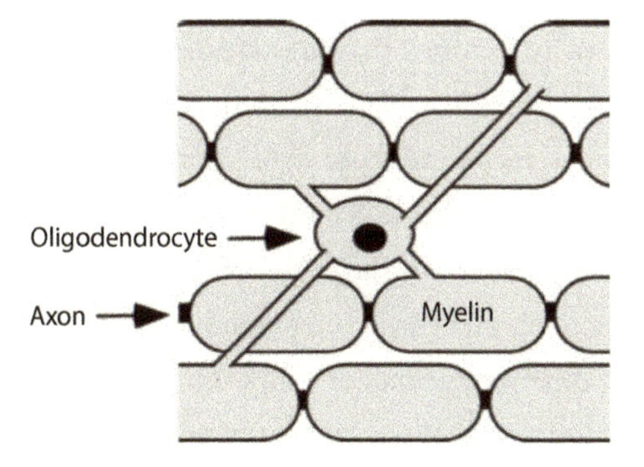

FIGURE 5.5 An oligodendrocyte (*middle*) supplying myelin to multiple different axons.

both myelin and oligodendrocytes become damaged in many different neurological diseases and during the course of normal aging, new oligodendrocytes can arise throughout life to replace damaged myelin.

In addition to the central nervous system (CNS), the axons of neurons outside the brain and spinal cord are also sometimes myelinated. This makes sense if you think about it: Myelin allows for high-speed signals between different brain areas, but the brain, via the spinal cord and special nerves (so-called cranial nerves), also connects to every part of our bodies, including the muscles and "sensors" that detect movement, touch, heat and cold, painful stimuli, and even sound. If the brain is signaling at high speeds (like a car on the autobahn) but the nerves connected to our brains and spinal cords signal at low speeds (like a car on a road at rush hour), none of the signals from the brain will be able to connect with the rest of the body in a reasonable amount of time, and we will not be able to properly move or sense our world (or practice, perform, or listen to music).

The axons of neurons that are outside the brain and spinal cord come together in cables that comprise the nerves throughout our bodies, including those in our arms and legs. The nerves and neurons outside the CNS make up the peripheral nervous system (PNS). Axons in the PNS get their myelin from cells that are really different from oligodendrocytes, called Schwann cells, after discoverer Theodor Schwann (1810–1882), who also found that the tissues of our bodies are made of cells. Unlike oligodendrocytes, which provide myelin for many different axons, Schwann cells in the PNS only make myelin for a single axon.

At birth, we humans have very little myelin. But soon after we are born, the number of oligodendrocytes in the brain and spinal cord increases dramatically and widespread myelination occurs in our first few years of childhood. It continues through adolescence and into early adulthood, finally slowing almost completely by the time we are in our twenties. Increasing the amount of myelin in our brains goes hand in hand with the formation of new circuits involved in all the things we do in our daily lives, including the development of motor, sensory, and cognitive abilities.

Studies in recent years have confirmed that we make new myelin even after our twenties and that learning, such as learning the fine motor skills involved in practicing music, can lead to new myelin formation.[12] One way this can happen is that the neural activity involved with learning may drive cells called oligodendrocyte progenitor cells (OPCs—the cells Larry studies in his lab) to divide and then differentiate into new oligodendrocytes that then replace old, damaged myelin or form myelin for new axons (figure 5.6).

This process is similar to what happens during neurogenesis, when neural stem cells become new neurons, the difference being that neural stem cells can become neurons, OPCs, and another type of glial cell, called astrocytes, while OPCs are limited to becoming new oligodendrocytes. They are the cells about to graduate from college to go off and become professional myelin makers.

Several studies have supported the notion that music practice per se leads to increased myelination and that the more you practice, the more you make myelin. These studies all use MRI-based

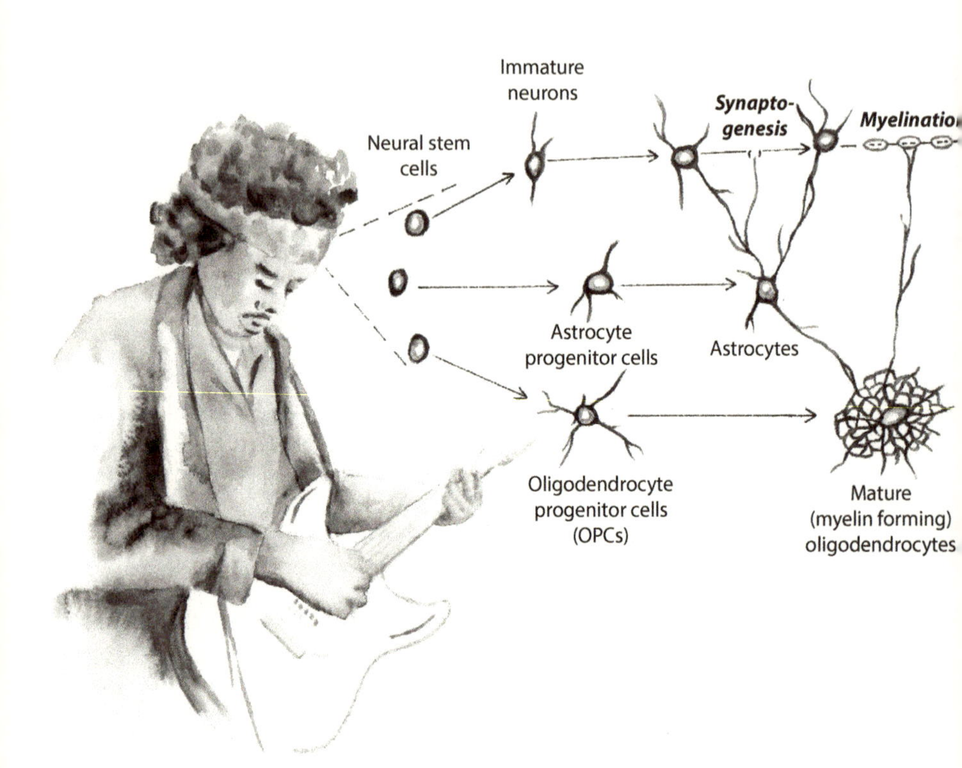

FIGURE 5.6 How music practice influences structural changes in the brain. Practicing an instrument engages motor, sensory, and cognitive challenges. Growing evidence shows that these types of combined challenges may induce neurogenesis, inducing some neural stem cells to become new neurons. These neurons mature and form synapses with existing neurons or other new neurons. Existing neurons can also generate new synapses. While neural stem cells can become OPCs or astrocyte progenitors, it is unclear if music practice generates them. Nonetheless, astrocyte progenitors will go on to form mature astrocytes that can interact with and regulate synapses, while OPCs become mature oligodendrocytes that myelinate axons, increasing the speed at which they conduct neural impulses. (Illustration by Susi Davis)

imaging techniques to follow the volumes of white matter in different brain regions. White matter is where huge collections of axons, most of which are myelinated, pass from one brain area to another. When you observe changes in the density or volume of white matter, it likely reflects increases in myelin.

One of the largest white matter areas is called the corpus callosum (which means "tough body"). It connects around 200 million axons from one hemisphere of the brain to the other. An early study supporting the idea that music practice can drive the formation of myelin examined the corpus callosum using an MRI technique in thirty professional musicians and thirty controls who matched in age, sex, and handedness.[13] The authors found that the corpus callosa of musicians who began training at an early age (at or before age seven) were significantly larger than those of musicians who started training later. Similar studies also found larger corpus callosa in musicians than in nonmusicians.[14]

Another study examined white matter structure in eight right-handed professional concert pianists, all around thirty years old, compared with white matter structure in eight male, age-matched nonmusicians.[15] The pianists in the group had all started playing piano at around the time they were six years old. The study estimated the amount of time the pianists spent practicing as young children, adolescents, and adults. They found that the number of practice hours during childhood correlated positively with increased measures of myelination, not only in the corpus callosum but also other white matter areas.

These findings indicate that myelination is most affected by music practice when we are younger, when myelin is still forming for the first time. However, before you become discouraged, another study examined the arcuate fasciculus, a tract of white matter that connects parts of the temporal lobe with parts of the frontal lobe, in a group of right-handed adults, aged eighteen to thirty, with limited or no musical training.[16] The subjects in this study had to learn specific finger movements using their left hands in response to either visual instructions or to music, and then perform this task during fMRI scanning (a similar challenge to learning to play a musical instrument). The study found that after only twenty minutes of training three times per week for four weeks, measures of myelination increased in the group moving their fingers in response to music. Therefore, even though the changes may be smaller than when we were young children, activities like practicing the piano or another instrument when we are older can lead to increases in myelination, improving the speed of nerve impulses between multiple different areas of our brains.

PRACTICE GIVES MUSICIANS SUPERPOWERS BEYOND PLAYING MUSIC

Given all of the cellular changes that occur with music practice, we might expect that some parts of musicians' brains other than white matter might be different from the brains of nonmusicians. Several studies have found differences in brain structure, including the sizes of some structures, in musicians. The brain areas that most consistently show increased volumes are the auditory cortex, which we discussed in the first movement, and motor areas, including the cerebellum.[17] These changes correlate with levels of music proficiency and the amount of time spent practicing.

If music practice leads to all these amazing structural changes in your brain, are those changes good for anything other than going from choppy to Chopin? Numerous studies have found that musicians have an enhanced ability to learn sensory and motor skills. For example, a study of fourteen professional pianists and sixteen nonmusicians found that pianists were far better at sensing the features of objects with their fingers (a process called tactile discrimination).[18] Tactile discrimination also becomes enhanced in people who are blind, giving them the ability to become skilled readers of braille. Musicians who started music training early in life also may learn complex motor skills unrelated to the instrument they play, better than music students who started later.[19]

The degree to which music practice enhances cognitive abilities remains controversial. It is difficult to control for the possibility that any given music student, before training, has different cognitive skills from those of nonmusic students. It is also difficult to control for all the ways that students are taught and how they learn. Nonetheless, a number of reports suggest that musicians who engage in substantial music practice perform better on certain cognitive tests, including memory.[20] Future studies will be needed to confirm these findings.

GREAT PRACTICE MAKES GREAT MUSICIANS

We have seen that successful practice involves a tremendous number of variables.

1. It requires a student who has certain traits, including motivation, passion, perseverance, and conscientiousness, and who engages consistently in the act of practicing music.
2. It requires a strategy to learn, guided by a teacher or devised by the students themselves, that promotes improvement and maintains the motivation to get better.
3. Finally, it involves the ability to engage multiple sensory functions (vision, touch, hearing) that integrate with motor and cognitive areas of the brain. With effort, practice leads to physical changes in the brain that are linked to improved nervous system function.

Once you have practiced to the point of generating new myelinated neural circuits that form neural networks, you are ready for the next step in the process of making music: performance.

Sixth Movement

HOW YOUR BRAIN PERFORMS MUSIC

YOU HAVE spent many hours practicing, converting the photons bouncing off the written music in front of you (or projecting from a screen) into movements of your hands and feet, or your mouth and vocal cords if you're singing. Or perhaps you have bypassed written music and your brain has worked out the music because you "play by ear." In any case, these movements you make result in changes in the air molecules your brain perceives as the music imagined in the brain of a composer. Over time, that practice has made your brain generate new circuits and alter old circuits until you are able to play the piece with ease.

Now you are ready to perform! Or are you?

Maybe you plan to perform just for yourself, or maybe for an audience. Perhaps you will do a solo performance to a small audience consisting of your music teacher and a group of other students nervously awaiting their turn (except for that one student—you know the one we mean—who can't wait to show off how they nail it every single time). Or perhaps you are performing in a nightclub in front of a larger group of people who are talkative and, in some cases, maybe more than a little drunk. Or maybe you are in a band or orchestra, playing in front of hundreds or even thousands of people in a concert hall or broadcasting studio. Or maybe you're alone in front of a microphone in a recording studio. No matter where you play, a performance is the culmination of all the time you have spent practicing, training your brain to play a specific piece of music on your instrument.

However, when the time for performance comes, it will seem nothing like practicing. Your performance will be transformed by how you are perceived by your audience (if you have one) and how you perceive their reception of your music. Performances may be altered in response to the venue and your understanding of the music based on your hours of practice. Performance, therefore, involves some different activities compared to your practice.

Our survey respondents shed some light on the differences between practice and performance:

- "When I perform, I am expressing a deep heartfelt sense of how the music has connected personally with me, while simultaneously being willing to connect with others, exposing my vulnerability to the possibility that they may not feel the same sense of connection with the music, and thus reject it and by association, me." (Dale Kardos, musician)

- "Practicing is figuring out how to do it, how to make the sounds that I want to make, working on the technical. It's working out the emotions of the piece. But performing feels like letting go of the technical and engaging in sharing the music, like telling a story." (Susannah Mars, actress and singer)
- "Performing involves an audience. I try to interact with the audience. Do I have their attention? What are they responding to? Where am I losing them? How could I perform this piece differently? How can I make slight changes as I perform in response to the audience? One time I was performing and the audience gradually started singing along on the chorus, so I added a couple of extra choruses at the end in response to their enthusiasm. I hadn't planned to do this. It just came naturally." (Chet Boddy, singer-songwriter)
- "Performing is communication, interacting through rhythm, melody, harmony, and improvisation—a cooperative process of sharing through mutual entrainment to sound and spirit." (Jeff Sweeney, musician)

These assertions illustrate several aspects of performance that differ from practice.

1. Most obviously, you want to know how the practiced piece may be perceived by the audience. Do they like it? Are they indifferent to it? Do they hate it?
2. The desire to share the music and the willingness to alter the performance are based on audience responses.

3. Finally, while some people might drag their feet about practicing, or even dread it, that doesn't come close to the level of trepidation that performing in front of an audience may invoke. Even if you are excited to finally share the effort of learning a piece, playing in front of an audience can be a fearful experience.

We will consider these aspects of performance and the neuroscience behind them.

THE MORE YOU (AND YOUR AUDIENCE) KNOW, THE BETTER YOU PLAY

Most performers feel a particular exuberance after the hours of deliberate practice pay off in a successful performance. Your music is being perceived by the brains of your audience, and, if all goes well, they are moved to sadness, joy, tenderness, longing, elation, awe, or some combination of several emotions.

So how does your brain, as a practiced performer, influence the brains of people who are listening to you? An important point to consider is that composers are highly dependent on performers, not to photocopy the original performance of a given piece of music, but to perform a piece appropriately, according to their own understanding. Performing publicly becomes an honor as well as a responsibility to combine the composer's voice with that of the performer. Thus, before performing a piece of music, performers should consider the background of the piece, its purpose, its composer, its style, and its context, and then represent all of it as best they can. As a result, performance can sometimes be enhanced by providing context for a piece of music prior to playing the music. When performers learn the meaning of a piece or the intent of a composer while practicing, they may find that sharing that intent (or, at least, their own understanding of that intent) will help an audience appreciate how the piece is being played or sung, thus connecting the brains of a performer and audience member before the first note is heard. The same is true for small groups or large orchestras. Explaining the purpose of the musical presentation can be helpful to the hearers, which is why orchestras include notes in their programs. Of course, some members of a given audience may be familiar with the program and come with their own expectations of what they should hear based on previous performances or recordings.

Things can go south rather quickly when the audience expects something other than what they receive. Even performers with powerful enough intent and cultural influence to engineer dramatic changes to their musicality will often find a frosty reception when an audience is unprepared for what they are hearing. In one famous example, the people attending the 1913 premiere of Igor Stravinsky's *The Rite of Spring*—now considered one of the most influential classical works of the twentieth century—were so scandalized by the avant-garde music and ballet that they broke out into what has been described as a near riot, including hissing and mocking the performers throughout the performance. Fifty years later, Bob Dylan, one of the top performers at the peak of the folk music movement, shocked the Newport Folk Festival on July 24, 1965, when he plugged in his electric guitar and played his iconic "Maggie's Farm." Though later accounts claim that there was applause mixed in with the audible booing, there is no question that Dylan going electric was initially not well received, and his angry fans harassed him throughout his tour. (And then everyone copied him and folk rock became popular.) This disconnect in expectations can be painful for both performers and listeners. Even artists with clear visions like Stravinsky and Dylan, who from a historical perspective appear to have been destined to initiate these cultural shifts, were upset by their audiences' hostility. From these anecdotes it appears that the bigger the challenge to expectations, the more time an audience might need to understand a new set of expectations and learn to trust where the composers are taking them. Conversely, perhaps the 1924 premiere of George Gershwin's *Rhapsody in Blue* was such a success with its audience (though not so much with the critics) because the piece was presented as "an experiment in modern music." In this case, expectations were set and met.

WHERE YOU PLAY AFFECTS HOW YOU PLAY

In addition to understanding *what* you are playing, it can be helpful to understand *where* you are performing. How those vibrating air molecules created by your instrument travel through a space and enter the ears of an audience is influenced by temperature, humidity, and, of course, if you are playing indoors, what many refer to as "the acoustics" of the space. The vibrating air molecules generated from a voice or instrument are altered depending on the shape and size of the room, the number of people in it, whether the floor is carpeted, and what materials cover the walls.

Amplification can also alter your performance, leaving much of how it sounds up to a sound engineer and his or her equipment (who might just be you and your amplifier). These variables affect both how the audience perceives your performance and how you hear it, influencing your own perceptions of the sounds you are making. Your piece might be entirely different in performance than in practice, perhaps requiring unanticipated adjustments in how you perform.

Here's the ideal performance space:

1. Throughout the space, you can hear yourself and everyone else who is performing.
2. Your audience can clearly hear the music, with proper balance between instruments and proper tonal balance (allowing optimal projection) for each voice and instrument.
3. There should be no extraneous sounds that detract from the performance (such as crinkling candy wrappers, excessive coughing, talking on phones, outside construction or traffic, or noisy neighbors).
4. The way sound persists in the space is appropriate for the music style. When a sound is reflected off walls, floors, objects, and even people in a space (reverberation), numerous reflections build up and then decay as those objects absorb sound.

Performance spaces with these conditions are hard to find, so be prepared to adjust your performance based on where you play. And try not to let those crinkling candy wrappers get to you.

YOU MOVE THE AUDIENCE AND THE AUDIENCE MOVES YOU

In addition to understanding how where you play affects your performance, be prepared to adjust your performance in response to who is listening. An audience may sit in silence during a performance, taking in every note that you play or sing. But audience members may also start to tap their feet or clap their hands. They may sing along. These sorts of responses may energize your performance. However, audience members may also talk to one another or get up and walk away, which would likely have a very different effect on how you perform.

A number of studies have shown that performers and their audiences have similar brain activity in their frontal, temporal, and parietal lobes during the course of a performance.[1] A recent study, using a technique called *functional near-infrared spectroscopy* (fNIRS), examined the brain activities of people who were drumming and of partners who were listening to the drumming.[2] Drumming has often been used to synchronize group activities (such as drum circles, dancing, and marching) and to facilitate listeners' emotional connection with events.[3] One might expect, therefore, that solo drumming would impact the emotional responses of both listener and performer. In this study, subjects (the drummers and the listeners) wore a cap with sensors that indicated roughly which areas of the brain were changing their activity during the drumming. This test was far less invasive than the fMRI studies mentioned earlier, but also less sensitive. The study found that those listening to drumming had increased activity in the area between the temporal and parietal lobes (the temporoparietal junction) in the right hemisphere, an area that is involved in paying attention to and analyzing events and that therefore affects social interactions. In contrast, drummers showed increased activity in their sensorimotor cortex while they were drumming.

Another study, also using fNIRS, examined the brain activities of a violinist and sixteen listeners.[4] Brain activity was measured in both the violinist and the listeners during live performances and while watching a video of the performance. The study found that brain activity in the violinist and the listeners was synchronized and that this link was strongest when the listener liked the music. These findings suggest that both performers and perceivers of a performance synchronize their brains to the music but that the popularity of a piece of music may influence this connection.

NEURON SEE, NEURON DO

How is this level of brain synchronization in performers and listeners achieved? In some brain areas, it is possible that groups of cells called *mirror neurons* may play a role. Studies in monkeys have demonstrated that certain groups of neurons, found mostly in areas of the brain that control movement, fire both when a monkey performs some activity like picking up a ball but also when the monkey observes a ball being picked up.[5] It has been proposed that this relationship between actual movement by an individual and perceived movement by others helps us understand the intentions of those we are watching.[6] So, when we see someone reach for our coffee cup, these neurons are essentially going through the motions of reaching for that coffee, presumably telling other parts of our brain that there is great danger of someone else drinking our coffee. Both singing and instrumental music involve complex movements, so when a performer is playing music, neurons involved in these movements are activated in the brains of both the performer and, at the same time, the audience. The activation of these mirror neurons may help a listener understand not only how the movement of a musician is related to the music being performed, but also the meaning of the music.[7]

What is less clear is how this synchronization and other signals in the brains of performers and their audiences are affected by both subtle audience responses (small head movements, smiles) and overt audience responses (foot tapping, singing along, cheering, or—gulp!—booing), and how the audience is affected by the movements of a performer.

Audience members frequently engage in movements when listening to live music. For example, one study using motion-capture devices recorded the head movements of audiences at a live concert featuring Canadian rock musician Ian Fletcher Thornley and at a separate event where an audience listened to recordings of the same songs. The live concert audience members demonstrated significantly more head movement, especially among self-reported fans of the music, than the audience listening to recorded music.[8] Such movement could be induced by the music itself or by the movements of performers.

As mentioned, audience responses can lead to a number of adjustments by performers, such as playing a part of the music a second or third time, improvising, or even changing the tempo. How the audience perceives what the performer is doing in response to the performance may similarly influence how they enjoy the music. Another interesting question is how responses by one group of audience members affect responses by other groups, which in turn can influence the performance (figure 6.1). For example, if you are performing and, among a

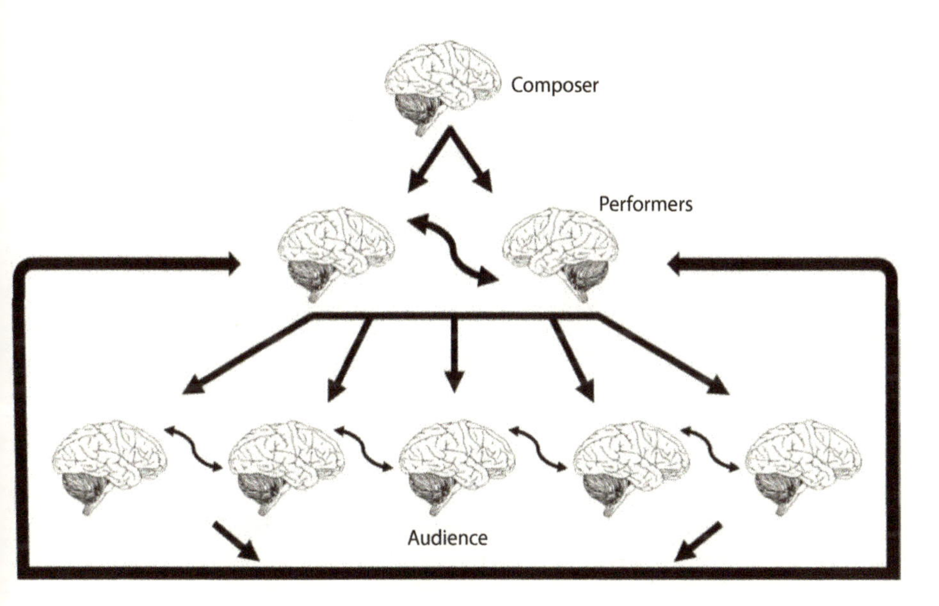

FIGURE 6.1 All the brains involved in musical performance. Musicians, alone or together, perform the music created by the brain of the composer. The brains of multiple performers may influence one another's performances. The brains of audience members respond to the music they hear in combination with what they know or what was presented to them before the performance. Their responses may influence one another's responses, which in turn may influence how the performers perform.

group of quiet listeners, several start to clap along to the music, then others in the audience will often start clapping. On the other hand, if the performance is not going smoothly, people in the audience may get up and leave or otherwise show their displeasure, encouraging others to do the same. Both outcomes may, in turn, influence how you or members of your group play, and they could disrupt the synchrony of brain activity between performers and listeners, leading to alterations in your performance and how much your audience enjoys it.

WHEN THE AUDIENCE IS A MICROPHONE

Performing in a recording studio is obviously a very different experience from performing for an audience. In a recording session, your main audience is microphones. As singer-songwriter Erin Cline put it in our survey:

> I have always found recording to be completely different than performing. For me, music is communication, and I like to see the faces of my audience so that I can see how they are responding. Recording is less enjoyable in this regard, and has never interested me as much as performing live. It is also very easy to become overly concerned with getting things "perfect," and as a result self-consciousness can become a barrier to a good performance.

In contrast, composer and musician John Smith reported that when he is performing for an audience, "There's a lot of internal dialogue going on: 'I'm going to play a whole-tone-half-tone scale here starting on the One chord.'" But in the studio, "A recording session is an orderly thing. Things happen in layers, and the picture emerges slowly."

Another big difference between performance and recording is that in the studio you are concerned with getting the sound just right, and you have the opportunity to make multiple takes to achieve that goal (depending on your budget). But there is very little communication, and all the aspects of a live performance are absent. Even if you are recording with a group, other performers might be separated by barriers or may be playing in separate glass enclosures. You might spend most of your energy trying to achieve the sound you practiced instead of adapting your performance to the space and a live audience.

MUSICIANS THAT PLAY TOGETHER, STAY TOGETHER (UNTIL THE END OF THE SONG)

Many performers report an incredible sense of freedom when performing *with* others on stage. Something magical happens when people perform together, communicating musically in a unified way. Whether singing in a choir or playing in a band or orchestra, the act of performing in a group is quite distinct from the experience of going it alone. Vocalist Joellen Sweeney, for example, said in our survey:

> Singing with a choir is one of my greatest joys in music making. Sometimes the adjustments I make to balance and

> blend with others are conscious—in rehearsal, they often are—but by the time we perform, it's more instinctual and intuitive. I love the feeling when a whole room of forty-plus singers is right in tune with the conductor, awake to their every minute motion. Sometimes I get too excited by the music, and the joyful, about-to-cry feeling makes it hard to sing.

Professional jazz pianist Bill Mays conveyed similar thoughts: "I am constantly hearing what the others are playing, even more so than I am listening to myself; responding to the general mood of the piece, the arc of the others' solos. . . ." Finally, John Smith wrote, "Lately the best part of playing live is the interaction with my fellow musicians. We all love each other as people, and we've known each other for so long. Some of us have played together for thirty years. When we're having fun, the audience can feel it. I get teary-eyed too often, when our band is on stage and everything is going right, and the audience is digging it. It's embarrassing."

When participants within a musical group care about the others and their respective roles in a performance, everyone performs with more meaning. "In short," as Andrew Stiller puts it, "music constitutes a powerful social glue. *Homo sapiens*, a social animal descended from a line of ancestral social species stretching back to the beginning of the Age of Mammals, thrives best when behaving cooperatively. Music helps to ensure that cooperation—indeed, must play an important role in that regard, or there would have been no need to evolve such a unique form of emotional communication."[9]

Interestingly, how musicians feel about one another may actually be driven by the act of performing together. Some unique things happen in our brains when we sing or play music in a group. Several studies have found that singing or playing music in a group leads to the release of endorphins, which are molecules that relieve stress and pain. For example, one study examined a community choir that met in both small groups (between 20 and 80 singers) or all together as a "megachoir" (232 singers). The singers from each small group and in the megachoir were tested for changes in their levels of social bonding and for pain thresholds (to reflect endorphin release). The authors reported that endorphin release and feelings of inclusion, connectivity, and positivity all increased during rehearsals for both groups.[10] Similar effects may be experienced when performing solo, but it appears that the larger the group, the greater the effect.[11] In the choir study, the large choir demonstrated a greater increase in social closeness than the smaller choirs. Thus, the larger the musical group in which you perform, the greater the impact on connectivity between the performers.

As mentioned, when we play music with others, interactions between the musicians can influence the way that everyone plays and how an audience perceives the performance (figure 6.1). Audience members may be certain of a group's excellent performance, yet if the relationships between the musicians are tense and competitive or otherwise exhibiting strife within the group, listeners may sense that friction, lessening the power of the otherwise impressive music making. The audience may not even perceive what they're picking up until they experience the contrast by hearing a group that is not as excellent but whose members demonstrate joy and freedom. In other words, a significant contribution to the quality of your performance will depend on the quality of the relationship you have with your group.

DIFFERENT KINDS OF MUSIC DEMAND DIFFERENT KINDS OF PERFORMANCES

Once you have the right place to play with a great audience (or adapt to whatever conditions you are faced with), you need to consider how the type of music you are performing will influence how you perform. This consideration is partly related to audience expectations, as we discussed earlier, but it is also partly related to the demands of different musical genres. For example, when Wu-Tang Clan performs "Protect Ya Neck," the energy and approach to the performance will be very different from the energy and approach of Neha Kakkar when she sings "Cheez Badi." And both of these performances will differ in approach and style from how Renée Fleming performs "O mio babbino caro."

The notion that our brains process different genres of music differently is supported by studies showing that different genres of music can influence different types of brain activity. For example, in the fMRI study with Sting, one task he was asked to do was to both listen to and imagine pieces of music from different genres (including "So What" by Miles Davis, "Mack the Knife" by Bobby Darin, "Satisfaction" by the Rolling Stones, and "The Great Gate of Kiev" by Mussorgsky). Interestingly, when listening or imagining these different pieces of music, the differences in which areas of his brain changed their activity appeared to be linked to the genre of the music. A more recent study of twenty-eight subjects without musical training examined changes in the auditory cortex after listening to instrumental fragments of classical, reggaetón, electronic, and folk music. The study found that the auditory cortex of the subjects showed different patterns and degrees of activation depending on which genre they listened to, and that reggaetón music had the strongest effects in not only the auditory cortex but also in motor-related areas.[12]

But do such changes occur in the brains of performers depending on the genre of music they perform? One study examined this question in terms of how different styles of music influence stage fright. A group of 239 students at German music schools were asked about their degrees of stage fright, and the data were examined with regard to the genres of music each student played. Interestingly, musicians between the ages of seven and sixteen who played primarily classical music reported the highest levels of stage fright, which was lower in older students. The students between the ages of seven and eleven who played primarily popular music reported much lower anxiety on stage, but older students tended to have higher levels of anxiety when performing.[13] Although the reasons behind these differences are unclear, they are consistent with the idea that musicians approach the performance of different music genres in different ways. We will discuss stage fright in more detail later.

It is possible that part of the reason that the German music students experienced different patterns of stage fright depending on whether they were performing classical or popular music had to do with the degree to which they had to stick to written music. Much the way composing and improvising differ in terms of their approaches and the ways our brains engage in each activity, performing from a score (written music) or a predetermined arrangement offers different challenges during a performance than jamming or improvising. When performing established music, you can rely on the composition's strength because it is already there and your role, as we discussed when practicing written music, is to interpret it meaningfully. But, of course, the audience expects that you will perform that music in a specific way. When your performance involves improvisation, the weight is more on you, for the performance has the potential to be anything from surprisingly spectacular to remarkably dull, depending on what you put into the performance. But you are also freer in how you play.

While the genre of the music you play may dictate specific aspects of your performance, at the end of the day how you practice for that performance will still be key to a successful experience for you and your audience. Both composer-driven music and performer-driven music require full concentration. Regardless of musical styles and the origins of the music, if you are technically, aurally, and intellectually prepared, what is needed during the actual performance is your full attention.

TRY NOT TO BE AFRAID—AT LEAST NOT VERY AFRAID

For reasons that still elude him, Dennis became a sensation at age seven playing the marimba. So he was on the stage repeatedly as a child performer, on television, radio, and with large live audiences. Having makeup applied to his face, having pictures taken, seeing himself in the newspapers, walking onto a stage, receiving applause, and being publicly recognized and appreciated became a normal way of life for an otherwise somewhat quiet, introspective boy. At a young age Dennis was acclimated to a way of life based on performing for others.

Larry, too, grew up enjoying performing for an audience. According to members of his family, he was a total ham at a very young age, playing piano for anyone who would listen. Later, he would participate in science fairs and science competitions, giving presentations to audiences about his projects. To this day Dennis and Larry love any kind of staged production.

When it comes to performance, Dennis's and Larry's stories are more the exception than the rule. For many, the thought of performing music in front of an audience is, at the very least, perceived as stressful. For others it is outright terrifying. Here it is, the time to share the results of preparation through practicing, but the overwhelming fear of others listening to you perform stymies a potentially joyous experience.

In our survey, we found that different people have different strategies to overcome these fears. For some, centering on the music, not the audience, is the answer. But for others, different interventions are required:

- "Stage fright is a result of two issues: self-absorption and lack of preparation. If I know myself and my purpose and my abilities, I should not be concerned about what the audience thinks of me. If I have prepared well, I can be confident." (Kelly Ballard, vocalist)

- "I have been solo performing since I was twelve. I still get nervous, sometimes to the point of panic attacks. I've tried everything, but a phobia is a phobia. After talking with my doctor, who is also a musician, I resorted to beta blockers and can finally experience the joy of singing at my full potential." (Peggy Schwarz, opera and folk singer)
- "The way I manage my brain when performing is simple: I make myself recall that everyone present wants to hear what I have to say musically, wants to hear a conversation between the musicians, and wants to be transported in some way; and since I know from experience that I'm able to play at that level, I build up my internal expectations on the spot and essentially say, 'You can do this, and they'll love it.'" (Cameron McMinn, jazz instrumentalist)

Stage fright is our brain's way of reacting to frightening situations. We learn to fear things that put us in situations where we are threatened. In most cases, this is a good thing. If we hadn't learned to fear danger, we, as a species, probably wouldn't be here. But not everything that we fear should be feared, and we often need to train our brains to see threats as challenges instead.

What is happening in our brains when we experience stage fright? On several different occasions when Larry was in school, the teacher or professor always explained this response using the example of wandering into a cave and encountering a vicious bear (like the one illustrated in figure 6.2). You immediately respond by either stopping in your tracks (freezing), confronting (fighting) the bear (usually a really bad idea), or running away (fleeing). This is called the freeze, fight, or flight response and involves a

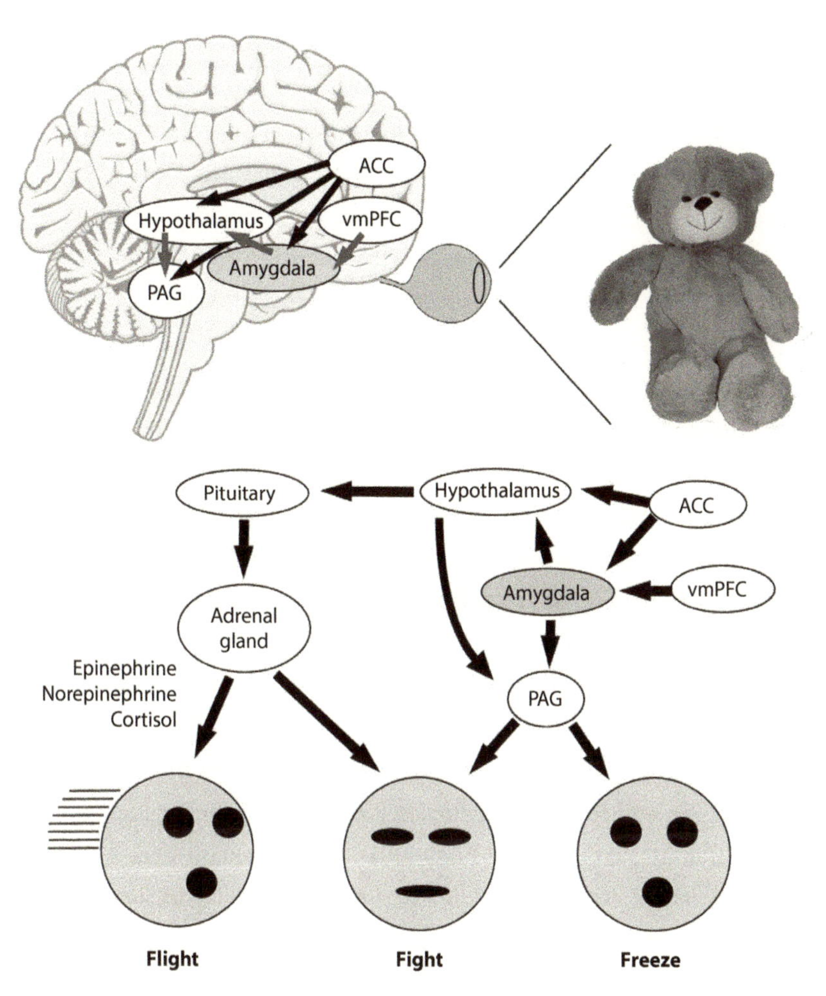

FIGURE 6.2 The circuitry of the freeze-fight-flight response—the basis of stage fright. Finding threatening bears or threatening audiences activates your amygdala and transmits signals to your periaqueductal gray area (PAG) and hypothalamus, which can cause you to freeze. But signals from the hypothalamus also signal to the pituitary gland and sympathetic nervous system, leading to the release of cortisol from the adrenal glands as well as other chemical messengers that prepare your body to fight (or push through a performance) or flee by getting your muscles, heart rate, blood pressure, and energy usage optimized for what comes next. Your anterior cingulate cortex (ACC) and ventromedial prefrontal cortex (vmPFC), found in the frontal lobe, regulate this response, helping you to assess whether you really need to run away or can go ahead and perform.

specific circuit involving the brain, the pituitary gland (in a place called the *sella turcica* or "Turkish saddle," a bone-enclosed space in the base of the skull, below the brain and behind the bridge of your nose), and the adrenal glands (found on top of your kidneys).

When you encounter something fearful, it activates a part of your brain called the amygdala. The amygdala has many functions, but one has to do with reacting to threats. The amygdala makes connections with many different brain areas, including a place involved in the freezing response (the periaqueductal gray area, PAG) and the part of the brain involved in releasing hormones and regulating emotional responses (hypothalamus) (figure 6.2). Signals from these areas alone can cause us to freeze in the cave or on stage.

However, in response to threats that activate the amygdala, the hypothalamus also sends signals via another circuit involving the sympathetic nervous system (which is *not* involved in causing you to feel sympathy, but rather in sending signals to different parts of your body so it is prepared to deal with threats). These signals go to a part of the adrenal glands called the medulla, which releases two neurotransmitters (epinephrine and norepinephrine) into the bloodstream. Together, these chemicals increase blood flow to muscles, increase blood output from the heart, help manage blood sugar levels, and increase blood pressure. The hypothalamus also sends a hormone (called *corticotropin-releasing factor*) to the pituitary gland, which in turn releases *adrenocorticotropic hormone* (ACTH) into the bloodstream. ACTH causes another part of the adrenal gland, called the cortex, to release a number of chemical messengers, including the stress hormone cortisol. During the flight or fight response, cortisol increases sugars (glucose) in the bloodstream and enhances your brain's use of glucose. All together, these processes prepare your body to run from the bear (or stage) or fight the bear (or the music you are about to perform).

The freeze-fight-flight response is the reason that our hearts pound and that we break out in a sweat in anticipation of a performance. Thankfully, with experience and strategies like the ones already quoted, we can overcome fear of performance and perform the music that we have practiced. This is in part because we have the ability to regulate the freeze-fight-flight response using our frontal lobes. In particular, areas called the *anterior cingulate cortex* (ACC) and the *ventromedial prefrontal cortex* (vmPFC) can help assess how much any given activity is an actual threat and regulate the amygdala's response (figure 6.2).

These and other areas of the brain can be trained to overcome fears in a number of ways. For example, if we continuously perform and our experiences are not negative, our brains will stop seeing performance as a threat, leading to less activation of the threat responses in the amygdala. This adjustment could come from external stimuli, such as consistent applause when we perform, or from focusing on our own gratification with performing the music that we have practiced. Furthermore, if we find performance to be rewarding (for example, if every time we play we are given a cookie or, better, money), we develop strong positive associations with performance that will also help us overcome stage fright. In either case, our heart might still pound and we might still break out in a sweat, but we'll be prepared to face our audience (and any bears therein) and to enjoy the experience of our performance.

Seventh Movement

HOW YOUR BRAIN LISTENS TO MUSIC

WHAT HAPPENS when your brain listens to music? How are the vibrations of air molecules created by music playing detected by music-specific centers in the human auditory cortex and sorted out in our nervous systems? What happens to signals in those music-specific areas of the auditory cortex that allow us to perceive music? And how does processing music affect us?

To get to this last question, we asked our survey respondents how listening to music affects them. The answers were remarkably diverse:

- "In taking time to listen to music, you discover things you've never noticed before. In taking time, you hear the beauty, complexity, and give honor to the person who took the time to compose the music." (Andrea Kahler, music lover and novelist)

- Listening to music is about "riding along with the song, looking at the composer in the eye and appreciating his work, anticipating what's next, being surprised and delighted, drifting away to dream about my own life and wishes, then coming back to the music and continuing the journey." (Wayne Harrel, playwright, lyricist, and scriptwriter)
- "I become totally focused, usually, to the exclusion of other things going on, like conversation. I have to be very careful when I am driving and listening!" (Shirley Brendlinger, pianist and music teacher)

Interestingly, many of the musicians who took our survey indicated that listening to music caused them to either imagine the experience of playing or singing what they were hearing or to analyze the music. As saxophonist and educator Tina Richerson put it, "When I listen with purpose, I am dissecting and learning."

These answers indicate that people with different musical backgrounds listen to music in different ways, all of which start with how our brains process and interpret what we are hearing.

SOUND'S JOURNEY TO THE AUDITORY CORTEX—AND BEYOND!

One of the brain's most incredible features is how it processes sensory information (like touch, taste, sight, smell, and hearing). In the case of sound, different sets of neurons fire in response to different types of sounds. Recall from the first movement that specific groups of cells may fire in the auditory cortex in response to music but not to other sounds. Remarkably, even before information about a sound gets to the auditory cortex, it is already organized so that sounds of different frequencies stimulate neurons in different places. That means that different musical notes are interpreted by the brain in different places.

This sound-to-neuron organization is called tonotopy (from Greek *tono* = frequency and *topos* = place). Tonotopy starts in the cochlea, the snail-like structure in the inner ear (fig. 7–1). Lining the cochlea is a structure called the basilar membrane, from which neurons transmit signals to different groups of cells in a part of the brainstem called the cochlear nucleus.

As we discussed in the first movement, small hair-like structures, called stereocilia, line the basilar membrane (which is in a portion of the cochlea called the Organ of Corti) and vibrate at different frequencies due to variations in thickness and width along the length of the membrane (figure 7.1). The cells associated with each of these stereocilia transmit information from different regions of the basilar membrane and encode frequency tonotopically. This tonotopy then continues through the vestibulocochlear nerve to the cochlear nucleus, the superior olivary complex, the inferior colliculus, the medial geniculate nucleus, and then all the way to the auditory cortex (you can review this pathway from figure 1.4).

Remarkably, using fMRI, it is possible to actually see different sets of neurons in the auditory cortex of someone listening to music "light up" when different tones are played, enough to predict what notes are being heard just by watching which cells are turning on.[1]

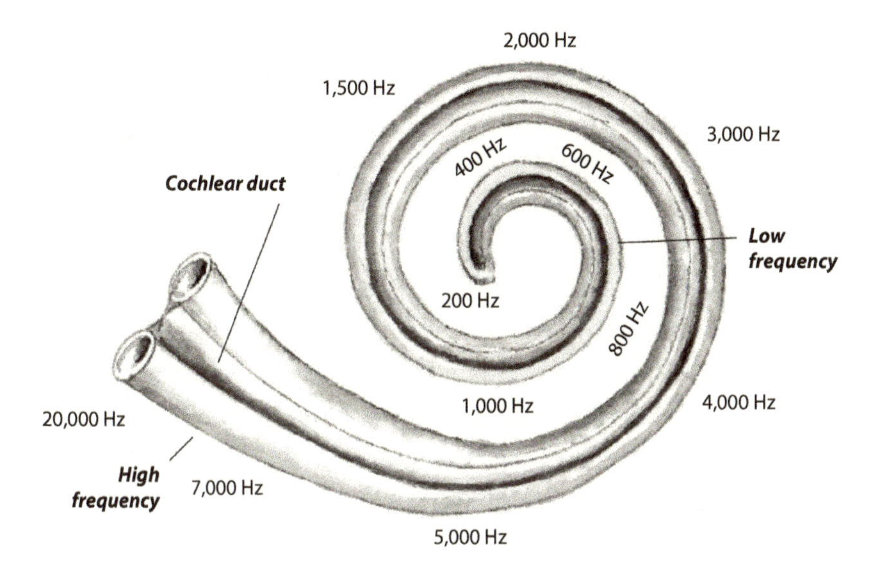

FIGURE 7.1 Tonotopic representation of sound in the Organ of Corti—the portion of the cochlea that processes sounds that enter the ear. Different parts of the basilar membrane react to different frequencies and then transmit that information to specific cells along the auditory pathway. (Illustration by Susi Davis)

HOW WE CAN TELL WHERE MUSIC IS COMING FROM

Additional information about sound is processed before signals reach the auditory cortex. For example, the first relay station in the brainstem after signals leave the cochlear nucleus—the superior olivary complex—receives information from both ears and is involved in sorting out where sounds are coming from. These cells measure the timing and sound intensity differences between signals from the left and right ears to calculate sound angles.

The next major area in the circuit—the *inferior colliculus*—helps you relate the time when you hear sounds to where they are coming from, as well as their pitch. Think about how you perceive an approaching car with its radio blaring. You know because the volume of the music is increasing that the car is getting closer. You also can get a general idea of the direction the car is coming from. And, if you don't like the music, your inferior colliculus helps you know when to yell "Turn down that noise!"

These areas of the brain also process the loudness of the music coming from different locations during a concert, giving you that rich experience of hearing instruments across the stage playing together but at different volumes. For example, when listening to an orchestra, if you close your eyes and concentrate, you can discern the instruments' placements on stage, with the violins playing on one side while the brass are on another.

MOVING ON UP, OR DOWN, OR NOT AT ALL

As sound processing continues, the medial geniculate nucleus coordinates information from the inferior colliculus and other areas of the brain and then sends it all to the auditory cortex. Once the cells in the auditory cortex are activated, each in response to specific frequencies and other characteristics of sound that are recognized as music by "music cells," neurons in the primary auditory cortex send signals to different places in the brain, including other areas of the auditory cortex (the secondary auditory cortex, which surrounds the primary auditory cortex in the temporal lobe).

Among these areas in the secondary auditory cortex, one network of neurons deals with contour. Recall from the first movement that in a piece of music, a melody's contour involves pitch changes (figure 7.2):

- *Moving up:* Consecutive groups of notes have higher and higher pitches.
- *Straight:* The overall pattern of notes remains at the same or similar pitches (for example, check out György Ligeti's *Musica ricercata*, which is entirely based around a single note: A).
- *Moving down:* Consecutive groups of notes have lower and lower pitches.

The network of cells involved with contour take information from the areas of auditory cortex that have sorted out the different notes and then calculate the relationship between those notes over time to give us the sense that the music is moving up, staying straight, or moving down.[2]

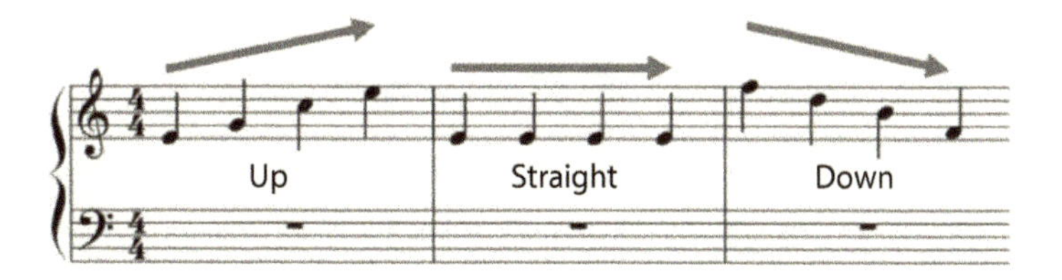

FIGURE 7.2 Contour is how we perceive the "direction" of a piece of music. It involves sorting out the relationships between pitches in musical sequences.

THE MINOR FALL AND THE MAJOR LIFT: SORTING OUT MINOR AND MAJOR CHORDS

Another function within areas of the secondary auditory cortex involves how we perceive different chords. For example, part of the auditory cortex (the superior temporal sulcus) appears to help distinguish major from minor chords.[3]

Remarkably, from there, major and minor chords are processed by different areas of the brain outside the auditory cortex, where they are assigned emotional meaning. For example, in Western music, minor keys are perceived as "serious" or "sad" and major keys are perceived as "bright" or "happy" (figure 7.3).[4] This is a remarkable response when you think about it: two or three notes played together for a brief period of time, without any other music, can make us think "that is a sad sound" or "that is a happy

FIGURE 7.3 Our brains assign emotional meaning to certain combinations of notes when played together (chords). In Western music, major chords (like C major) are "happy" or "bright," while changing a single note (in this case, the E to an E-flat) changes the chord to a minor chord that is thought of as "sad" or "serious."

sound." People around the world have this response, although the tones that illicit these emotions differ from one culture to another.

In a study of how the brain reacts to consonant chords (notes that sound "good" together, like middle C and the E and G above middle C, as in the opening chord of Billy Joel's "Piano Man"; figure 7.3), subjects were played consonant or dissonant chords (notes that sound "bad" together) in the minor and major keys, and their brains were analyzed using a method called *positron emission tomography* (PET). This method of measuring brain activity is different from the fMRI studies we discussed earlier. PET scanning, like fMRI, can be used to monitor blood flow in the brain as a measure of brain activity, but it uses tracer molecules that are injected into the subjects' bloodstreams. Although the approach is different, many of the caveats we mentioned for fMRI studies also apply to PET studies. Nonetheless, these authors reported that minor chords activated an area of the brain involved in reward and emotion processing (the right striatum), while major chords induced significant activity in an area important for integrating and making sense of sensory information from various parts of the brain (the left middle temporal gyrus).[5] These findings suggest the locations of pathways in the brain that contribute to a sense of happiness or sadness in response to certain stimuli, like music.

DON'T WORRY, BE HAPPY (OR SAD): HOW COMPOSERS MANIPULATE OUR EMOTIONS

Although major and minor chords by themselves can elicit "happy" or "sad" emotions, our emotional response to music that combines major and minor chords with certain tempos, lyrics, and melodies is more complex. For example, the emotional link to simple chords can have a significant and dynamic impact on the sentiments in lyrics. In some of his talks on the neuroscience of music, Larry, working with singer, pianist, and songwriter Naomi LaViolette, demonstrates this point using Leonard Cohen's widely known and beloved song "Hallelujah." Larry introduces the song as an example of how music can influence the meaning of lyrics, and then he plays an upbeat ragtime, with mostly major chords, while Naomi sings Cohen's lyrics. The audience laughs, but it also finds that the lyrics have far less emotional impact than when sung to the original slow-paced music with several minor chords.

Songwriters take advantage of this effect all the time to highlight their lyrics' emotional meaning. A study of guitar tablatures (a form of writing down music for guitar) examined the relationship between major and minor chords paired with lyrics and what is called emotional valence: In psychology, emotions considered to have a negative valence include anger and fear, while emotions with positive valence include joy. The study found that major chords are associated with higher-valence lyrics,[6] which is consistent with previous studies showing that major chords evoke more positive emotional responses than minor chords.[7] Thus, in Western music, pairing sad words or phrases with minor chords, and happy words or phrases with major chords, is an effective way to manipulate an audience's feelings. Doing the opposite can, at the very least, muddle the meaning of the words but can also bring complexity and beauty to the message in the music.

Manipulative composers appear to have been around for a long time. Music was an important part of ancient Greek culture. Although today we read works such as Homer's *Iliad* and *Odyssey*, these texts were meant to be sung with instrumental accompaniment. Surviving texts from many works include detailed information about the notes, scales, effects, and instruments to be used, and the meter of each piece can be deduced from the poetry (for example, the dactylic hexameter of Homer and other epic poetry). Armand D'Angour, a professor of classics at Oxford University, has recently recreated the sounds of ancient Greek music using original texts, music notation, and replicated instruments such as the aulos, which consists of two double-reed pipes played simultaneously by a single performer.[8] Professor D'Angour has organized concerts based on some of these texts, reviving music that has not been heard for over 2,500 years. His work reveals that the music then, like now, uses major and minor tones and changes in meter to highlight the lyrics' emotional intent. Simple changes in tones elicited emotional responses in the brains of ancient Greeks just as they do today, indicating that our recognition of the emotional value of these tones has been part of how our brains respond to music deep into antiquity.

ONLY A FOOL AND YOUR FRONTAL LOBE PLAY BY THE RULES

Cells in the auditory cortex connect with various centers in the frontal lobe. These areas in the frontal lobe transform information from the auditory cortex and elsewhere into what we experience as the various parts of music. They also monitor the rules and expectations for music, reacting when the rules are broken.

- *Timbre:* One example is timbre, which, as we discussed earlier, is the nature of sound that tells you its source (for example, a specific musical instrument or voice). If your timbre circuitry is intact, you can distinguish a middle C played for exactly one second on a piano from a middle C played on any other instrument. You could probably even distinguish a middle C played on different brands of the same instrument (like a Steinway versus a Baldwin piano). Although some aspects of timbre are already distinguished by the auditory cortex,[9] our ability to perceive the source of any note as coming from a specific instrument or voice is processed in parts of the frontal lobe.[10] So, as with other sounds, much of the basic information about the sounds associated with music is processed in the different parts of the auditory cortex and then sent elsewhere to be interpreted as what we perceive as the different parts of music, including timbre.
- *Patterns:* The frontal lobe, among other areas, also keeps track of the patterns of music that we have learned over the course of our lives. As a result, when we hear a series of chords or notes, we predict the next notes in the piece. For example, when someone plays a scale like C, D, E, F, we expect the next note to be G (figure 7.4). But if the next note played is G-flat (which is also F-sharp), our frontal lobe reacts, letting us know that this note violates the rules. It goes against every expectation we have for what comes next in that sequence.[11]

This is also how we perceive music as consonant or dissonant, as already discussed, because we have expectations for patterns of chords and other aspects of melodies. Interestingly, our tolerance for the degree to which such rules are broken may influence how much we like certain styles of music. (More on this question in the next movement.)

- *Modulation:* The frontal lobe is also involved in detecting key changes in a piece of music. Music tends to establish a "tonal center," where musicians reach a consensus on a particular key. But music often changes to another key or keys (which is called modulation). Our brains react, sometimes with heightened emotions, if a piece of music is played through a few times in one key (based on a selected set of notes) and then moves to a different set of notes. This is yet another way that composers manipulate our emotions (not that there's anything wrong with that).

FIGURE 7.4 Our frontal lobes play a role in assessing how what we hear agrees with how we understand the rules of music and what we expect in musical patterns. The fifth note (gray) in the sequence (scale) shown here is G, which we expect in this scale to follow the F.

RHYTHM: TAPPING WITH YOUR CEREBELLUM TO THE BEAT

Humans experience rhythm in daily life as movement and patterns perceived by their brains. People relate to rhythm in varied ways. Whether being rocked or rocking out, humans respond reflexively with clapping, toe tapping, swaying, or even dancing.

Some experience rhythm in music without a noticeable beat, such as Gregorian chants, where rhythm flows freely, unmeasured, based largely on the rhythms of speech. In contrast, most popular Western music has a specific rhythm that is linked to the timing of the music such as 4/4 (also called common time), where each measure has 4 beats (figure 7.4). The rhythm of some music can be more complex, such as Dave Brubeck's "Take Five," which has a 5/4 time signature, where the beats in each measure can be thought of as being split into two uneven chunks.

Whether music flows without a sense of beats grouped into regular units or flows in a beat-centric way, our brains organize what they hear by perceiving patterns that we recognize as rhythm. But whereas a metronome's beat is mechanically produced, humans can add feeling to the elements they perform over the beat's framework. When musicians speak of "feel," they refer to a distinctive quality discernible in the performer's rhythmic patterns. For example, for a piece of music in 4/4 time, the first and third beats are typically played with greater emphasis. If the emphasis is on the second and the fourth beat, as is often the case in jazz and rock, the music feels "different" because of the consistent expectation of the first and third beats.

Without the normally expected one and three, two and four would not have their uniqueness.

To experience this effect, sing the simple tune "Mary Had a Little Lamb," clapping on one and three:

Mar—y had a litt—le lamb (rest)

<u>1</u> 2 <u>3</u> 4

Now try singing the tune again while clapping on two and four:

Mar—y had a litt—le lamb (rest)

1 <u>2</u> 3 <u>4</u>

The feel changes. The rhythmic orientation greatly affects the style. It's as basic as <u>1</u> 2 <u>3</u> 4 or 1 <u>2</u> 3 <u>4</u>.

Once a rhythmic pattern has been established in a piece of music, breaking the pattern by any alteration gets our attention, much the way an unexpected note that breaks the rules in an expected tonal pattern jars our brains (for example, the scale in figure 7.4). If a piece features a stress on beats two and four, then suddenly switches to one and three, we experience an emotional shift.

Such an unexpected change to an established pattern is referred to as syncopation, which is common, for example, in the works of ragtime composers like Scott Joplin. Thwarting expectations with stresses where they would not normally occur creates energy. While too many surprises can result in chaos and incomprehensibility for some listeners, some people actually like it that way. If the unexpectedly changed pattern becomes the rule, not an exception, then it becomes a fresh basis for expectation, felt or perceived as normal. The ebb and flow of expectations help create a sense of forward motion, of moving toward some goal or endpoint.

So where does the brain process rhythm? It turns out that the cerebellum, which we discussed earlier with regards to its other roles in music processing, may play a crucial role in the detection and perception of rhythm in music. Although it makes up only about 10 percent of the total volume of your brain, the cerebellum contains more than 50 percent of the total number of neurons in the human brain, giving it tremendous processing abilities. A variety of studies have suggested that the cerebellum is essential for understanding changes in tempo when listening to the beats in a piece of music.[12] It is also involved in determining small differences in the length of time between beats.[13]

Of course, the cerebellum does not act alone in how you perceive rhythm. It is highly connected with many other areas of the brain, such as the groups of neurons in the basal ganglia. These cells have a wide range of functions, including controlling motor activity and emotions. In a study on how the cerebellum and basal ganglia function to interpret rhythm, investigators brought together groups of subjects that either had no brain damage or who had damage to either their cerebellum or basal ganglia. The subjects with damage in the cerebellum had problems processing fast rhythms. They could not keep up with quick beats. In contrast, the subjects with basal ganglia damage had problems with complex rhythms.[14] So, for example, they could keep up with music in 4/4 time, but a piece like "Take Five" would be problematic. Such rhythms require more complex patterning of the beat, suggesting that beat complexity requires the basal ganglia.

The fact that the cerebellum and basal ganglia also have connections to areas of the cortex that control movement likely explains our instinctive response to rhythms, which causes us to move to the beat of the music we hear even at a young age.[15] Rhythm, once detected and maintained, activates neural circuits involved in motor processing, suggesting that there could be direct circuits connecting rhythm centers and movement centers in our brains.[16] Indeed, Fernández-Miranda and colleagues identified direct connections from the auditory cortex to motor areas (the so-called primary motor cortex and premotor cortex), suggesting even more direct responses to music in our motor center.[17] In other words, we are fully wired from early childhood to move to the beats we hear.[18] It is no wonder that music and dance have evolved together in many cultures.[19]

MUSIC CAN MAKE YOU SO EMOTIONAL

Although simple chords can be perceived as "happy" or "sad," more complete pieces of music can significantly alter mood and evoke a far greater range of emotions. You have probably found that some songs or compositions can lift you up when you are feeling down, while other music can make you feel sad. This is largely because the areas of the brain that process music are both directly and indirectly connected with parts of the brain that influence our emotions.

These emotion centers—collectively called the limbic system—include these parts of the brain:

- *Amygdala:* Brain-imaging studies have found, for example, that the perception of joyful music is linked to one group of neurons in the amygdala (an area referred to as the superficial nuclei) that are involved with understanding the social significance of information coming from other areas of the brain.[20]
- *Hippocampus:* Connected to both the auditory cortex and the amygdala, this has been implicated in regulating emotions like nostalgia, wonder, tenderness, and peacefulness.[21]
- *Nucleus accumbens:* An area of the brain we haven't discussed before, this has multiple functions, including reward signaling (for example, reinforcing addictive behaviors). The nucleus accumbens has been linked to signaling intense feelings of pleasure in response to music.[22]
- Other parts of the amygdala, especially the right laterobasal amygdala: These are involved in our responses to joyful music but also to sad or unpleasant music.[23] The right laterobasal amygdala has been implicated in processing both negative and positive associations. When we are watching a movie and sad music starts playing before a scene, the right laterobasal amygdala may be responsible for making us anticipate that something sad is about to happen. The fact that the auditory cortex has so many direct and indirect connections to the limbic system highlights the fact that our brains are wired to have emotional

responses to sound, and it may explain the power of music in altering our emotions.

Music that is recognized as happy or sad may trigger different brain networks than music that promotes other emotions. Furthermore, music may elicit emotions through pathways that are distinct from pathways elicited by other stimuli. An fMRI study of 102 subjects examined which brain areas responded to music from movie soundtracks and other music rated as happy, sad, fearful, or tender and compared the results with responses to film clips that contained various emotional and nonemotional content.[24] Although there were some overlaps, emotions evoked by films and music involved different brain networks. The authors speculate that this is because films depict emotions linked to real-life situations while music is more abstract. Interestingly, activity in the auditory cortex and primary motor cortex could discriminate the different types of emotion, such that the patterns of activity could predict whether a subject was listening to happy, sad, fearful, or tender music. This remarkable finding could help reveal the fundamental ways that humans experience emotions in response to music and other stimuli.

I WILL REMEMBER YOU; WILL YOU REMEMBER ME?

In addition to its role in emotional responses to music, the limbic system plays an important role in learning and memory. Emotions enhance memory formation. In our daily lives, our brains place many events into short-term memory, but most are forgotten. Events that mean something to us are assigned emotional value and stored in our long-term memory. Because music evokes strong emotions, the limbic system can be involved in forming long-term memories in response to pieces of music or in relation to episodes and information (such as lyrics or events) associated with particular music. These circuits therefore have also been linked to the ways that music can bring up certain memories and the emotional responses to those memories.[25] For example, if you happen to be listening to "Hey Ya!" by Outkast when you come to realize that you are falling in love with the person you are with, hearing "Hey Ya!" later in your life will elicit the memory of that moment (for better or for worse). So it's understandable that many romantic couples claim a certain song as "theirs."

MUSICIANS WHO LISTEN TO MUSIC ARE PLAYING THE MUSIC THEY LISTEN TO—IN THEIR BRAINS

As we mentioned at the beginning of this movement, many of our musically trained survey respondents indicated that when they listen to music, they sometimes analyze what they are hearing. Some even envision playing the instruments that they hear or go through the motions of playing (like solo air guitar). Recall the concept of mirror neurons that we discussed in the sixth movement; these neurons in brain areas that deal with movement become activated both in the brains of performers and in the brains of audience members who listen to those performers. Mirror neurons may also be activated in musicians who are just listening to either live music performed by other musicians or to recorded music. For example, musicians who just imagine playing a piece of music activate multiple areas of their brains, including areas involved in motor functions.[26] Motor areas were also reported to be activated in pianists who watched silent videos of hands playing a keyboard.[27] Thus, just the perception or imagination of music can trigger a musician's brain to fire neurons involved in the performance of that music.

DOES JUST LISTENING TO MUSIC MAKE YOU SMARTER? SPOILER ALERT—NO

As you can see from the studies we have reviewed here, just listening to music engages many parts of your brain that integrate elements of music, emotions, and memory. So given the demands of music listening on the human brain, you might expect that music listening itself might somehow improve brain function.

As we have discussed, listening to music can have many effects on our brains, including altering our mood, which can itself have other positive effects on the brain, such as reducing stress responses.[28] In addition, music listening in early childhood can have a number of positive effects on brain development, including how our brains process and react to sounds in general.[29]

But can simply listening to music make you smarter? In 1993, a study by Rauscher and colleagues examined how undergraduates (who will often volunteer for such studies in exchange for pizza) scored on spatial reasoning tasks (like solving mazes) after sitting in silence or after listening to the first movement *(Allegro con spirito)* of the Mozart Sonata for Two Pianos in D Major.[30] The study found that those subjects who listened to this piece of music from Mozart scored better than the control group who did not listen to the music. The study was exaggerated in the popular media and led to claims of a "Mozart effect": the idea that simply exposing children to the music of Mozart would make them smarter. Companies sold recordings of Mozart to parents, claiming that they would improve the intelligence of their kids. At one point, the governor of Georgia, Zell Miller, issued a proclamation that every mother of a newborn should get a free compact disc of Mozart's music to stimulate her children's brain development.

Sadly, not only was the public interpretation of the Raucher et al. study wrong, but the actual study was flawed. Multiple subsequent studies failed to see a significant effect of Mozart or other music on spatial or other cognitive tasks.[31] Some studies demonstrated trends toward improvements in some tasks, but these improvements were likely linked to the fact that music can increase arousal. In other words, if you are sitting in silence, your brain will not be particularly active compared to when your brain is listening to bright, upbeat music, which can affect the activity of your cerebral cortex. For this reason, subjects who are exposed to arousing music are more likely to perform better on spatial tests such as mazes, largely as a result of being more alert.[32]

This effect of music on arousal explains why many people wake up to music, work out in the gym to music, and play music while they are doing other activities (such as homework). Larry, for example, typically listens to music when he is writing grants or papers (and he listened to music when he wrote portions of this book). But music selection is important when added to work that requires writing and similar brain functions. If the music is too emotional, including sad music or music with lyrics that make Larry want to sing along or dance, his attention turns entirely to the music and he isn't very productive.

The reason lies in the fact that, as already discussed, joyful music and sad or tense music activate different centers in our brains. One study demonstrated that attention improved in subjects who were exposed to pleasant and arousing music.[33] The authors found that joyful music was associated with greater activation in the centers of the brain that regulate attention (including parts of the posterior parietal lobe and the right frontal lobe). In contrast, exposure to tense music enhanced activity in more posterior areas of the brain, while sad music was associated with increased activity in the visual cortex. In other words, sad and tense music can elicit emotions that might distract your attention, while joyful music activates attention centers. So, the next time you have waited until the last minute to write a report or complete some project, put on some joyful music (without lyrics) at low volume and get to work.

Eighth Movement

WHY YOUR BRAIN LIKES MUSIC

WE ARE delighted that you have made it this far in this book (unless, of course, you cheated and skipped ahead). As we have seen, composing, practicing, performing, and listening to music engage our brains in remarkable and highly demanding ways. We made the argument that music exists for many reasons, from enhancing human communication to bonding people together. These arguments all support the idea that music gave us an evolutionary advantage by enhancing our ability to interact and live in groups.

But a simple answer for why humans make music is that we really, *really* like it. Music is a significant source of pleasure for most of us, and some of our most basic behaviors are driven by pleasure. In this movement, we explore the nature of pleasure, the different ways that music brings us pleasure, and the cellular and molecular mechanisms that influence how we experience pleasure. Then we explore possible reasons that sad songs can be pleasurable for some people and how our brains come to prefer certain genres or pieces of music. Finally, we'll travel to the Amazon to explore if some of the ways we respond to music are universal to the human brain.

NO PROFIT GROWS WHERE NO PLEASURE IS TAKEN—W. SHAKESPEARE

Coming back to the ancient Greeks, Socrates and Plato gave a lot of thought to pleasure and even went so far as to define different forms of pleasure:

- *physical* (which were basically eating and sex),
- *aesthetic* (admiring beauty, including music), and
- *ideal* (intellectual pleasures, like sitting around all day thinking about what pleasure really is).

Experiencing pleasure (*hedonia*) was defined as a state of cheerfulness linked to liking something or some state of being. And when you had a perfect amount of hedonia in your life, you were said to have achieved a state of *eudaimonia*—a "life well lived"—basically, a sense of well-being.

Today, what Socrates and Plato called "physical" pleasures are referred to as "fundamental" pleasures—examples of adaptive responses, which means they evolved to help our species survive. Why do we need to experience pleasure when we eat and have sex? Well, if you think about it, both eating and sex are pretty strange things to do. Eating involves biting into and tearing apart plants or animals, then pulverizing them by chewing and swallowing the resulting bits and pieces of whatever we consider food.

Imagine if food didn't bring a sense of pleasure when you ate it. In fact, some people lose the ability to experience pleasure, a condition called anhedonia. Many of them lose weight because they lose the pleasure associated with eating and eat less. In the case of sex, again, if you think about all the things you do when having sex, they are not necessarily activities that people would want to engage in if they didn't know how pleasurable they can be. ("You put what, where?") But, because we derive pleasure from sex, we overlook that strangeness. Again, in the example of people with anhedonia, they do not find sex pleasurable and often avoid it.

Wiring our brains to enjoy food and sex makes perfect sense. If our species were uninterested in food or sex, we wouldn't be here wondering why our brains like music so much. But music is not generally considered a fundamental pleasure; it is grouped among "higher-order pleasures" along with art, money, altruism, and spirituality or transcendental pleasures. Music is, nonetheless, considered by many to be nearly as important as the fundamental pleasures. People rank music among the things that bring them the most pleasure, usually above money, art, and even food.[1] So perhaps because music does play an important role in sustaining human beings, our brains are wired to find it as enjoyable as fundamental pleasures.

PEOPLE FIND PLEASURE IN MUSIC IN MANY DIFFERENT WAYS

In a separate survey from the one underlying the other themes in this book, we asked our respondents to anonymously share how music brings them pleasure. While we received an unsurprisingly wide variety of responses, this one in particular caught our attention: "Just imagine the pleasure from eating dark chocolate *while* making love and listening to Bach . . . now that's a dose of pleasure!"

But even without dark chocolate and sex, music elicits strong feelings of pleasure in many different contexts. Many respondents spoke about how listening to a musical performance impacts them in emotionally intense ways. One person, a twenty-year-old woman, described listening to a violin-piano duo by a husband and wife who are both professional musicians. She found their beautiful music compelling, unrelenting. As they pressed forward into a most intense musical passage, the violin in a high register, the music was so dynamic that she grabbed her friend's hand. They both squeezed as hard as possible just to have some sort of physical outlet for the powerful beauty that they were experiencing. She truly thought her body would explode. She didn't want the sound to end, but she needed it to end because she didn't think she could handle more of its intensity.

Another memorable musical experience one of our respondents described happened at a jazz club when he was in college. When one of the players took a solo, our respondent was completely drawn in. He felt that he was on a ride down a track, as on a roller coaster ride, where he could see what was coming, but it was thrilling and surprising at the same time. He felt like crying, with his body tense and moving with the music in a gripping, emotional way, as people do when swaying with laughter or rocking while sobbing. There was anticipation and tension, like watching a movie scene unfold, interspersed with moments of relief. When it was over, he felt happily exhausted, as if he'd just done a hard day's work. And that feeling has stuck with him for over twelve years now.

Several of our respondents wrote about how *performing* music brings them pleasure and how music performance has helped them get through doing other tasks. One respondent, for example, indicated that she absolutely loves to sing. She draws great pleasure when performing with a group, particularly after working on a piece in rehearsal for as long as it takes to achieve the choir's optimum performance.

That said, she also wrote that singing alone or with just one other person was just as pleasurable and could even make an onerous task much more bearable. When her younger sister and she were growing up, their kitchen was not equipped with an automatic dishwasher. As their mother put it, "I have two dishwashers!" So she and her sister were responsible for washing and cleaning up everything, including after parties. To make the task go by more quickly, the sisters sang, but they didn't sing songs. They did improvisational vocalizing, singing random counterpoint vocal lines. They would go up, down, in, out, and around each other's notes. They were mostly interested in just hearing each other and responding. They would probably vocalize for at least a good half hour at a stretch. The singing gave them much more pleasure in doing their domestic task than they would have experienced without creating those marvelous, purposeless, strings of notes. Similar to the old "whistle while you work" idea, it certainly brought them joy.

Earlier, we discussed how singing or playing an instrument in groups can produce endorphins, which can relieve pain. But playing music alone may also influence pain intensity. One respondent, in her seventies, practices piano one hour each day. She said she struggles with pain in her back, but she noticed, particularly when learning a new Chopin piece, that when she goes to bed at night or even if she is just sitting during the day, the new music will go through her head and make her feel better. By becoming more aware of the positions of her body, hands, and fingers and how those positions relate to the sound of the music, all while hearing the rhythms as intensely as the melody, she can appreciate both playing and listening. She admits that she can enjoy lots of music without knowing about how it is put together but that understanding how music is made brings her even more pleasure.

For many of us, listening to music or attending a concert can similarly relieve (or cause us to relive) emotional pain, including the pain associated with ending a relationship. One of our respondents shared how music helped him following his first big romantic rejection. He was enjoying the company of several close friends at the Homowo Festival at the Washington Park Amphitheater in Portland, Oregon (where Dennis and Larry live), which was headlined by Obo Addy and Kukrudu—an eight-member African jazz group that relied on a mix of European and African instruments and that was headed by Addy, an outstanding and celebrated Ghanaian percussionist and singer. While enjoying his friends, our respondent was still feeling emotional pain from his recent loss. His days were painful. He went to bed in pain. He awoke in pain. Here he was with these friends looking at the stage and hearing the music and feeling apathetic (and pathetic). Then, suddenly, the music infected him and joy began to spread through his body. He couldn't help it: he stood up and danced for the next hour with a rush of happiness that, although temporary, provided a powerful respite from his pain.

Similarly, one of our respondents described suffering with opiate addiction, during which the climb out of that pit was three steps forward and two back. He shared about a difficult period when, four days into having quit "cold turkey," his body was responding to his abstinence with profound unease and dissatisfaction with life (a condition called dysphoria), physical discomfort, and cravings for more opioids. Normally he loved and listened to music, but when his use of drugs was heaviest, he stopped engaging with the things that usually brought him joy, including listening to music. For some reason, at this extremely low point in his life, experiencing this physical and emotional pain, something inspired him to turn on the radio. A certain upbeat and energetic pop rock song was on that he knew and liked. As he lay on his couch, he was amazed at how much joy he experienced listening to the song. He could feel a warm comfort move through his body. He remembers thinking, "It's going to be okay," and he had a sense of hope. He guesses that this moment jump-started his brain into making its own "feel-good juices" again. These many years later, he still uses music as therapy during times of emotional discomfort, remembering that couch moment that taught him how music can uplift him.

Recall our discussion in the last movement on the link between listening to music and memory. One of our respondents shared how she powerfully associates music with her childhood memories, and two in particular. When she was growing up, her family often played classical music. She and her sister would put on their skating skirts with white satin lining and dance around the living room, convinced that they were as beautiful as any ballerina. Their father loved American tenor and Hollywood actor Mario Lanza. She can still hear Lanza's lyrics "Guardian angels around my bed, keeping watch in the night" in his tenor voice, which bring strong, fond memories about her family. Another soundtrack in her young life was a set of 45-rpm records called *Rusty in Orchestraville*, in which Rusty is a young boy who hates practicing piano and has a dream in which the conductor of Orchestraville takes him around to meet all the instruments. Our respondent and her sister listened to that record over and over, learning the sounds of the instruments and how they contributed to the orchestra. She reported that decades later the two sisters listened to the song and their memories came flooding back. She wrote that she could see herself in the living room of the original family home enthralled by the music.

Another respondent gained pleasure from music's ability to convey stories. For him, music bears many resemblances to the structure of story, taking place chronologically over time, comprising a series of tense moments followed by resolution, and typically having some sort of climax near the end. When he listens to a film's soundtrack, he not only listens to the music, but he also recalls each scene, moment by moment, synchronized with the soundtrack. He relives the story, almost as if he were watching the film over again. He believes there is no better example of music's storytelling ability than *Peter and the Wolf* by Sergei Prokofiev. The symphony tells a story, assigning each character an instrument and a theme. The details of the story are recounted by a narrator, but the bulk of the drama is created by the music. The story is told, in part, by the use and order of the themes. He cites John Powell's scores for the *How to Train Your Dragon* movies as contemporary examples of his assertion that music can help to tell stories in visceral ways, getting at truth that words and plot are unable to explain. He claims that the triumph, action, and exhilaration of riding a dragon are conveyed by the score alone much better than by the rest of the elements of the film. Clearly, music can hold emotional truth as capably as the plot, words, or images of a film.

This idea was supported by a respondent who wrote that, when taking voice lessons, one of her assignments had been to learn the Italian aria *Nina*. The melody was emotionally powerful to her. During a lesson when she was first learning this piece, she turned to her instructor and said, "I have no idea what I am singing but I'm certain someone has died." The voice teacher laughed and said, "Well, actually, Nina has been lying on her death bed for three days and appears to be close to death. Her lover is lamenting."

One young man spent time kayaking with friends at his favorite lake and brought his soprano sax with him. He often took it there because of the "magical" resonance created by the cliffs around the north end of the lake. On a midsummer evening, at a comfortable eighty degrees, he pulled out his horn and sent a few arpeggios out over the water. After a few minutes someone answered back on a flute. Whoever it was (he never located the person), this phantom flutist was a considerably better musician, because anything our respondent played was sent back to him perfectly. As the sun got low, the two musicians kept going and people came out on their porches along the edges of the lake to listen. This went on for about an hour, with the two trading off ideas and even at one point breaking into "trading fours" on a twelve-bar blues riff. Then, as it got dark, a chilly breeze came up. He ended a musical "sentence" with a pause, it was answered perfectly, and that was that. The listeners around the lake clapped, his friends paddled over to high-five him, and he will never know who the flutist was. Chances are the story is also being told to this day from the flutist's point of view.

Collectively, these responses indicate that people seek to derive pleasure through music in multiple, different ways. Consistent with that idea, David Huron, a professor at Ohio State University who focuses on music cognition and the psychology of music, has suggested that the pleasure we derive from music is linked to how we "use" it. In one study, he and his colleagues examined data from multiple theoretical and empirical studies on the function of music (for example, studies where subjects were just observed) and found that although communication and social connectivity were important, people reported overall that the most important effects of listening to music had to do with self-awareness, arousal, and mood regulation. In the category of self-awareness, people reported that listening to music helps them think about who they are, who they would like to be, and how to find their way in life. In the arousal and mood regulation category, people reported that music listening was an important way to take their minds off of reality, to help them change their mood, and to help with both relaxation and alertness. These findings reflect the many reasons that people listen to music and the many pleasurable ways that the brain reacts to it.[2]

MUSIC IS LIKE A DRUG

Overall, the ways that music influences pleasure are similar to the ways that our brains process other pleasurable experiences. For example, people who have damage to pleasure centers in their brains (for example, parts of the limbic system) stop finding music pleasurable, much the way they stop finding other things or activities pleasurable.[3] It is nonetheless remarkable how powerfully music can influence the cellular and chemical signals involved in pleasure.

Back in the third movement, we discussed the relationship between liking and wanting. Recall that when our brains experience something that they like (such as chocolate or great—or even not-so-great—sex), this liking response leads us to want more of the thing that we like. When we get it, we get a sense of reward, which gives us a sense of liking again (figure 8.1). A large part of the circuit involved in this liking-wanting loop is in the nucleus accumbens, which, as we mentioned in the previous movement, has been linked to signaling intense feelings of pleasure in response to music.

Liking involves neurochemical pathways that include the so-called mu (as in the Greek letter μ) opioid and cannabinoid receptor pathways. As the name suggests, mu-opioid receptors are proteins that recognize and cause neurons to respond to opium (specifically, morphine, which is the main psychoactive chemical in opium). When we experience something we like, our bodies release substances similar to morphine in our brains and activate our mu-opioid receptors. For example, when we eat chocolate, our bodies release an opioid called enkephalin

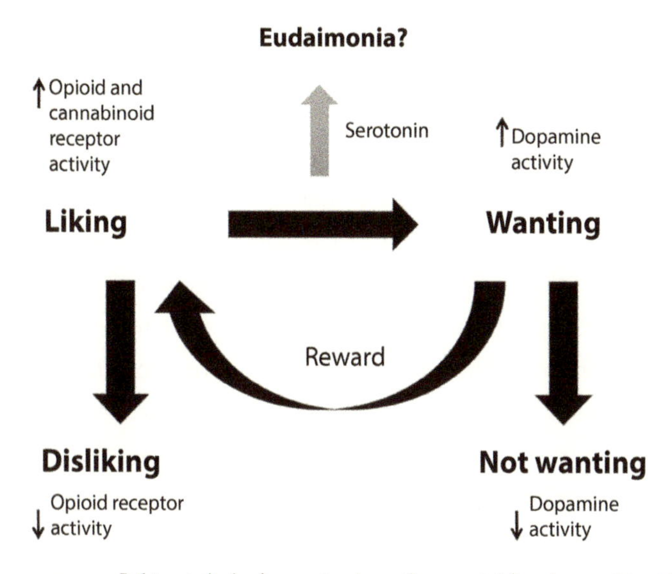

FIGURE 8.1 Liking is linked to activation of mu-opioid and cannabinoid receptors, while wanting is mostly linked to dopamine activity. When liking and wanting are in balance, other signals, such as the neurotransmitter serotonin, are released, giving us a sense of satisfaction. This may be the state of eudaimonia described by Plato and Socrates. When we have had enough of any pleasurable stimulus, opioid, cannabinoid, and dopamine receptors are reduced, reducing liking and wanting responses.

that stimulates the mu-opioid receptor on neurons, leading to pleasure, relaxation, and contentment.[4]

Because of these functions, drugs that activate mu-opioid receptors, including morphine, have long been used to relieve pain. These drugs do not directly impact the source of pain but rather lead you to care less about the pain because of the pleasurable effects that come with activating mu-opioid receptors. Activities or experiences that activate mu-opioid receptors may similarly reduce pain. For example, let's say you go out dancing with somebody that you really like. You're having a great time and feeling really good with this person. Then, suddenly, your partner steps on your foot. It hurts, but it doesn't hurt that much because you are enjoying your evening (which leads to the release of opioids that activate your mu-opioid receptors). Now consider a similar evening of dancing but with someone you don't get along with. You are *not* having a good time and can't wait for this date to end. You reluctantly agree to head out to the dance floor. When this person steps on your foot, the pain is much worse because your mu-opioid receptors are not activated and you are not experiencing pleasure that can mask the intense pain that this clumsy person has inflicted on you.

Much like the way we respond to chocolate or having a great time on a date, music that we like may similarly bring us pleasure through the activation of mu-opioid receptors. In one study, males and females (fifteen subjects in total) were given either a placebo or a drug that blocks mu-opioid receptors (naltrexone). The subjects then listened to two pieces of music that they had previously reported finding pleasurable and two pieces of music that they didn't feel one way or the other about ("neutral" music). After this listening session, the subjects were tested for how pleasurable they found the music using a number of measurements. Subjects treated with naltrexone no longer felt pleasure when listening to music that they normally found pleasurable.[5] Therefore, opioids made by cells in the brain (like enkephalin) may be responsible for why we like certain songs or even entire genres of music. They may also explain why the pleasure we experience when listening to certain music we like can reduce both emotional and physical pain, as mentioned by some of our survey respondents.

DOO BEE DOO BEE DOOBIE DO . . .

As already mentioned, cannabinoid receptors are also involved when we like something. These proteins, found at the surfaces of neurons, signal in response to chemicals known as cannabinoids. Among cannabinoids is what is most commonly known as THC (Δ9-tetrahydrocannabinol), which is found in marijuana (cannabis). When you smoke a joint (or eat pot edibles), THC is what makes you feel high.

It turns out that certain pleasurable experiences cause us to make our own cannabinoids, called endocannabinoids. The euphoric feeling that some runners experience (as opposed to the pain and exhaustion experienced by others) after a long run is partly due to increased circulating endocannabinoids in their bloodstreams. This "runner's high" is described as a sense of well-being and improved mood, and it can include reduced anxiety. Although it is unclear if listening to pleasurable music can similarly increase levels of endocannabinoids, one study of members of a choir (all female, with an average age of sixty-one) found that singing significantly increased circulating endocannabinoids.[6] Thus, while singing or playing instruments in a group affects the production of endorphins (as discussed in the seventh movement), singing may also impact the production of signals that influence mood and pleasure (and it's much cheaper than weed).

Although the role of endocannabinoids in our enjoyment of music is unclear, people often turn to using cannabinoids to enhance their musical experience. Subjects in several studies report that smoking or otherwise consuming weed enhances their appreciation of music.[7] Marijuana use is consistently high (no pun intended) among people who attend music festivals, nightclubs, and other music venues.[8] Given that music engages the same regions in the reward network of our brains influenced by cannabinoids, the association between cannabis and music performance may be partly attributable to shared effects on reward circuitry between drug use and music listening.[9]

MUSIC YOU LIKE MAKES YOU WANT TO HEAR MUSIC YOU LIKE

While liking involves one set of neurochemical signals, wanting is largely linked to another chemical called dopamine (figure 8.1). Dopamine is linked to your desire for some thing, some activity, or some person. It drives the desire to seek out things that you like. Dopamine is a neurotransmitter that is synthesized in various parts of the brain (it's one of the molecules that axons release into synapses and that are taken up by dendrites). In addition to its role in wanting, dopamine influences movement, sleep, mood, attention, and how we think about things.

Similar to the liking pathways, much of the dopamine reward system involves the nucleus accumbens. Using the PET scanning approach already discussed, one study examined what happens to dopamine when we listen to pleasurable music.[10] The study reported that when subjects listened to music that they had previously told the scientists they found pleasurable, dopamine was released in the nucleus accumbens and other brain areas. In other words, when we anticipate hearing a certain piece of favorite music, actually hearing it activates our reward circuitry. If you tell someone that you really like "Louie Louie" by the Kingsmen, and then they play it for you, you will get a strong dopamine rush in response to your wanting to hear it. And then you will like what you hear because hearing "Louie Louie" will lead to the activation of your mu-opioid receptors.

Dopamine may also be directly involved in our ability to enjoy the music that we like. The best evidence of this link comes from a study by Laura Ferreri and colleagues, who examined a group of twenty-seven male and female subjects who were given, in random order, a placebo, a drug that increases dopamine levels at synapses (levo-dopa), or a drug that blocks dopamine activity (risperidone), and then they were asked to listen to music.[11] The pleasurable (hedonic) responses to the music were assessed using physiological measures and ratings by the subjects; for example, they were asked how much money they would spend to buy the music.

The study found that increasing dopamine led to a greater pleasurable response to music, while blocking dopamine had the opposite effect. One caveat to these findings: risperidone can also block serotonin activity, which, as already mentioned, may be involved in that sense of well-being achieved when liking and wanting are in balance. Nonetheless, these and previous findings indicate that dopamine plays an important role not only in our desire to seek out the music we like but also in the pleasure we experience once we hear the music that we seek.

I (AND MUSIC) CAN'T MAKE YOU LOVE ME

While music drives the liking, wanting, and reward signals in the brains of many people, some do not find music at all pleasurable. On the one hand, there is a rare condition called *amusia*, which is the inability to recognize musical tones or to make them through singing or playing a musical instrument. Although some people with this condition are called "tone deaf," others have much more severe impairments in their ability to process music when they hear it.

People can be born with amusia (called congenital amusia), due to problems with neural development, or they can develop amusia as a result of brain injury from trauma or disease (like strokes). Studies of people with congenital amusia have suggested that they have structural deficits in the connections between different areas of the brain that are needed to process music. These structural problems can include a lack of myelin (making the connections between music-processing areas slower than they need to be) or changes in how the axons from neurons connect between brain regions.[12] So, people with amusia cannot enjoy music because they cannot properly perceive what music is.

On the other hand, there are people who have no problems perceiving music but still derive no joy from it. We described a condition called anhedonia in which people derive little or no pleasure from any experience. However, some people specifically do not find music pleasurable. They still find joy in everything else, such as paintings by famous artists and even nonmusical sounds (like a baby laughing or an audience applauding).[13] Such individuals are said to have "musical anhedonia."[14] Like people with amusia, these individuals can develop this lack of musical enjoyment as a result of brain damage. However, between 5 and 10 percent of people without brain damage throughout the world have musical anhedonia, suggesting that their brains developed in a way that prevents them from enjoying music. The degree to which musical anhedonia is due to genetics versus exposure to music and other stimuli in early childhood is unknown.

A number of studies have attempted to identify the differences in the brains of people with musical anhedonia compared to people who enjoy music. Recall that a major part of the brain involved in processing rewards is the nucleus accumbens. A study in the laboratory of Josep Marco-Pallarés at the University of Barcelona identified students who really liked music, generally liked music, or who were found to have music anhedonia.[15] The students with music anhedonia demonstrated far less activation of the nucleus accumbens when listening to music compared with subjects who enjoyed music but had the same levels of activation when engaging in a gambling task with a monetary reward. The students with musical anhedonia also showed reduced connectivity between their auditory cortex and the nucleus accumbens (as well as other parts of the brain's reward network).

These findings highlight the idea that music drives a type of reward in the brain that requires direct connections between the auditory cortex and reward centers. Indeed, a subsequent study found that a person with severe musical anhedonia had reduced connectivity in their white matter connecting auditory and reward centers (white matter is where most of the myelin-covered axons communicate between brain areas). Among control subjects, the degree of this connectivity correlated with the degree to which they liked music.[16] This and other studies support the idea that liking music in general requires very specific connections between our auditory cortex and our reward centers. The fact that people with amusia and musical anhedonia can still process other sounds raises the possibility that they may still benefit from exposure to music even if they cannot perceive or enjoy what they are hearing.

YOU MIGHT AS WELL FACE IT YOU'RE POSSIBLY ADDICTED TO MUSIC

At the opposite extreme of people who have musical anhedonia are people who cannot get enough music in their lives. Like music, drugs of abuse alter our emotions and influence the many pathways we have discussed that are involved in our experiencing pleasure when listening to music. Indeed, some have argued that it is possible to be addicted to music listening and music practice, although these notions are considered controversial.[17]

But how do we become addicted to something? One idea, for which there is significant support, suggests that drug abuse, at least, involves a state of excessive wanting without liking or at least with limited liking.[18] This so-called *incentive-sensitization theory* may apply to various "behavioral addictions" that do not involve drugs at all: eating, gambling, sex or pornography, Internet use, shopping addictions, and so on. But whether it applies to what some are calling "music addiction," where someone feels compelled to listen to a specific piece or genre of music or just music in general, is unclear. Nonetheless, music can be "used" as a coping mechanism for stress and anxiety, much the way drugs of abuse can be used to self-medicate as a coping mechanism,[19] and it is plausible that excessive music listening might distract someone from work and daily routines to the point of negatively impacting their life. It may not be technically addictive, but listening to or playing music to excess may have negative consequences, especially if used to mask an unresolved psychological challenge.

WHAT MAKES A SAD SONG SAD?

Aside from sad lyrics, music we perceive as sad tends to have certain characteristics that include, especially in Western music, chords or stretches of melody in minor keys, lower overall pitch, narrow pitch range, slower tempo, use of dull and dark voices (for example, timbres in vocal or instrumental music), softer and lower sound levels, and less energetic performance.[20]

SAD SONGS SAY SO MUCH

If we associate certain music or even certain chords with happiness and pleasure, and we seek out things and activities that make us happy, why are there so many sad songs? If you listen to the radio or any music-streaming applications on shuffle mode, you will find songs about lost love, death, or some other conflict. The songs may have sad lyrics combined with sad or uplifting melodies. For example, Eric Clapton's "Tears in Heaven," written with Will Jennings, is a memorial to Clapton's son Conor, who tragically fell to his death from a New York City apartment building at age four. While it is one of the saddest modern songs, it remains popular, in part, because of the hopeful-sounding melody.

Similarly, one of Larry's favorite songs is "Man of Constant Sorrow," an American folk song attributed to Dick Burnett, a partially blind fiddler from Kentucky, from around 1913. As the song's title suggests, it does not inspire happiness, and yet it has been incredibly popular for over a century, in part because the music has an upbeat melody and rhythm.

In contrast, some of the most enduring compositions have some of the famously saddest melodies. Samuel Barber's *Adagio for Strings*, for example, is among the most popular of all twentieth-century orchestral works. But the music is incredibly sad. It was played at the funerals of Albert Einstein and Princess Grace of Monaco, and over the radio at the announcement of U.S. president Franklin D. Roosevelt's death and following President John F. Kennedy's funeral. For many people, it's intertwined with the climactic scenes of David Lynch's *The Elephant Man* and Oliver Stone's *Platoon*. Also, the BBC Symphony Orchestra played it during the last night of the Proms in London, just days after the attacks on September 11, 2001.

Although everyone reacts differently to "sad" music, one model describes music as both sad and pleasurable when (1) we find it nonthreatening, (2) we find it aesthetically pleasing, and

(3) it produces psychological benefits, such as altering mood or promoting feelings of empathy from memories or reflections on past events.[21] So, when you listen to something like *Adagio for Strings*, you may find the music both comforting and beautiful, and it may remind you of someone, something, or some place, bringing both sadness and pleasure at the same time.

Do people feel genuine sadness when listening to sad music? Although there remains some debate about this question, queries of music listeners tend to indicate that what they experience is indeed true sadness, although more transient than the sadness they feel in response to events in their lives. Philosopher Jerrold Levinson has suggested that there are eight different benefits of feeling sadness when listening to sad music:[22]

(1) *Catharsis*, which can involve feeling the sadness of an event without experiencing the actual event, helping you release negative emotions
(2) *Apprehending expression*, where you develop an improved understanding of the emotions expressed in a piece of art or music
(3) *Savoring feeling*, the joy of feeling emotion in response to art
(4) *Understanding feeling*, which is recognizing and learning about your feelings as you experience the music
(5) *Emotional assurance*, where you recognize your own ability to feel something deeply
(6) *Emotional resolution*, knowing that your emotions and emotional responses to music and other stimuli can be regulated

(7) *Expressive potency*, the pleasure you experience when you express your feelings
(8) *Emotional communion*, that connection you feel with the composer, performers, or other listeners

In support of these concepts, studies in which participants were asked to provide their motives for listening to sad music have shown that people often cite similar benefits to the ones described by Levinson.[23]

While at least some people who like, play, and listen to sad music appear to derive these benefits from the experience—and it could be argued that these alone are enough of a reward—some questions remain about why sad music can bring us pleasure.

David Huron from Ohio State University and Jonna Vuoskoski from the University of Oslo have proposed what they call a *pleasurable compassion theory* to explain how sad music can inspire joy.[24] The idea behind this theory is that sad music sounds similar to the vocalizations that humans make while grieving or when they are feeling melancholy. When hearing these sounds, listeners who have a high level of empathic concern for others experience feelings of sympathy or compassion, which are related to altruistic acts (figure 8.2).

When we engage in different forms of altruism, we activate pleasure networks in our brains.[25] This makes sense from an evolutionary standpoint. Like eating and sex, altruism is an adaptive behavior that helps ensure our survival (or at least the survival of people we feel close to). Interestingly, people who like and react to sad music tend to be easily absorbed by stories in books and films; they are said to have a *trait empathy for*

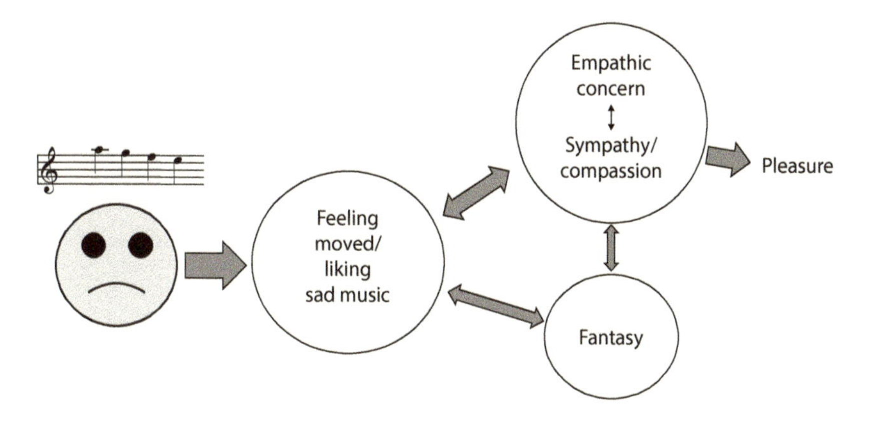

FIGURE 8.2 If you are moved by sad music and have high levels of empathic concern, you feel sympathy or compassion. These feelings are enhanced when people also have high-level empathy for fantasy. For such people, feeling sympathy or compassion, like altruism, activates pleasure networks in the brain.

fantasy.[26] As part of the pleasurable compassion theory, having fantasy as a trait can enhance the pleasure experienced when listening to sad songs.

I CAN'T BELIEVE THAT YOU ACTUALLY LIKE *THAT* SONG!

Aside from liking (or not liking) sad songs, we each have distinct preferences for the types of music that we listen to. Our brains are constantly making judgments about whether or not we like the music we hear. Why we like some genres of music, particular pieces within a genre, or even particular versions of a piece of music over others is unclear. It is likely that musical experience (what we are exposed to at different times during our lives), social interactions (what our friends and families listen to), age, values, and personality traits play a role.[27] For example, one study that examined personality traits found that people who are open to new experiences often prefer jazz, blues, folk, and classical music, while people who are extraverted and agreeable tend to prefer pop, soul, funk, electronic, soundtrack, religious, and dance music.[28]

Another study by Cambridge University psychologist and researcher David Greenberg and colleagues found a distinct correlation between brain types or thinking styles and musical preferences.[29] In particular, this study examined the roles of empathy and systemizing in determining what music people like. Empathy is the ability to sense, predict, and respond appropriately to the mental states of other people. People use empathy when they listen to music and when they perform it, which lead to physiological and emotional responses to the music.[30] Systemizing involves recognizing, predicting, and responding to the ways that systems work. It involves analyzing expectations and rules about systems, like those found in music (including scales and chord progressions, as discussed earlier).

The Cambridge study involved over four thousand subjects divided into three groups:

- *Empathizers:* Tend to focus on people's emotions and thoughts; prefer "mellow" music with gentle, warm, and sensual attributes and negative valence (depressing and sad) and emotional depth (poetic, relaxing, and thoughtful), such as rhythm and blues or soul, adult contemporary, and soft rock

- *Systemizers:* Tend to focus on rules and systems; prefer music that is strong, tense, and thrilling and that has positive valence (is animated) and cerebral depth (complexity), such as punk, heavy metal, and hard rock
- *Balanced group:* Part empathizers and part systemizers

One finding from this study indicated that empathy levels are linked to preferences for certain kinds of music, even specific music within genres. In fact, the level of empathy a person has predicts music preferences better than certain personality traits.

WHERE YOU ARE FROM MAY DETERMINE YOUR TASTE IN MUSIC

How people define consonance and dissonance varies across different cultures. Even in the same culture, what is consonant and dissonant may change over time or depending on a listener's musical experience. Josh McDermott and colleagues demonstrated this point in a study that compared responses to consonant and dissonant tones in people from Bolivia, the United States, and in the Tsimané people, a native Amazonian society with minimal exposure to Western culture.[31] Although the Tsimané subjects were able to tell the difference between various tones to the same degree as the subjects from Bolivia and the United States, they rated consonant and dissonant chords and vocal harmonies as equally pleasant, while the Bolivian and U.S. subjects tended to like consonant music and did not like hearing dissonant music.

The results indicate that preferences for consonance do not develop in people who have no experience with Western music. So, our responses to consonant and dissonant music do not reflect some innate bias to sound or exposure to harmonic natural sounds, but rather exposure to musical harmony.

Although the Tsimané people did not share the same responses to consonant and dissonant bits of music that were observed in Bolivian and U.S. subjects, a later study found that the Tsimané were as good as subjects exposed to Western music when they had to distinguish between tones played together.

In particular, Tsimané subjects and subjects in Boston, Massachusetts, were tested for their ability to distinguish when two tones played together sound like a single note (a phenomenon called *fusion*) or when they are clearly two different notes. Subjects heard two notes, played at the same time, separated by a particular musical interval (the distance in pitch between two notes). Tsimané subjects and Bostonians were equally likely to mistake combinations predicted to create fusion, and they were able to distinguish notes with intervals that did not create fusion.[32] These findings indicate that some aspects of musical perception are, indeed, universal.

LOVE (OF A SONG) IS A CURIOUS THING

We discussed how curiosity is important for music composition in the second movement, and how curiosity contributes to the motivation to learn music in the third movement. Curiosity may also contribute to music preference. As with other human behaviors, listening to music can be exploratory. We hear something new and become curious about the rest of the piece. The degree of curiosity about a piece or style of music can determine how long we engage with that music. This idea is supported by a study of users of the popular music-identifying application *Shazam*,[33] which can provide information about a piece of music when the listener samples a fragment of the music using a phone or computer. Users of this application hear a piece of music and become curious about the music, and then seek more information about it. The study found that the musical moments that drove curiosity tended to be highly salient (such as the opening chords of a song, the start of the chorus, or the onset of vocals). These findings suggest that our curiosity about a piece of music, and our desire to hear more, are driven by just a few seconds of music that grab our attention.

What is it about those few seconds that we find so appealing? Omigie and Ricci explored the possibility that it's all about how changes in music trigger our curiosity.[34] In their study, subjects were asked to rate their levels of curiosity about what will happen next while listening to particular pieces of music, their level of calmness, and the level of perceived changes in the music over time. The subjects first practiced this task while listening to a 40-second fragment of Herbie Hancock's "A Tribute to Someone." Then, they listened to short fragments of the adagio from Luigi Boccherini's Cello Concerto in D Major, published in 1770; the rondo from Igor Stravinsky's Concerto in D for Strings from 1946; an electronic dance music piece called Piku by the Chemical Brothers, released in 1997; and finally Dance the Devil Away, a piece of music by the group Outback that fuses Australian indigenous music with Western music. None of the subjects were familiar with any of these pieces of music before participating in the study. The subjects consistently reported increased curiosity, but not feelings of calm, whenever they perceived unexpected changes in the music they were hearing. When our expectations about what comes next in music are challenged, our curiosity and drive to hear more are piqued and we go on listening.

WE LIKE MUSIC BECAUSE WE NEED MUSIC

From the fMRI studies back in the second movement that demonstrated music-specific areas in the auditory cortex, to the findings of perceived musical fusion among the Tsimané people, it is clear that our brains are intricately wired for music and are shaped and reshaped throughout our lives by our musical experiences.

Learning to play an instrument or sing can drive the generation of new cells, new synapses, and new myelin in our brains as we combine fine and gross movements with the complex sensations of vision, touch, and hearing to generate music, either in isolation, in a group, or with an audience whose brains are affected by what we play and whose reactions affect how we play. Moreover, music, like food and sex, has the capacity to dramatically influence our emotions and to bring a tremendous level of pleasure to those who learn to play, perform, or listen to it, activating networks of neurons that reinforce our desire to play and learn more.

For some, music is a powerful reward that enhances life experience, with the capacity to bring people together and communicate stories for generations to come. Music is the most fundamental of the higher-order pleasures. In the words of the German philosopher, scholar, and writer Friedrich Nietzsche, during a holiday in the Swiss village of Segl Maria: "Wie wenig gehört zum Glücke! Der Ton eines Dudelsacks.—Ohne Musik wäre das Leben ein Irrthum. Der Deutsche denkt sich selbst Gott liedersingend," which translates to "How little is required for pleasure! The sound of a bagpipe. Without music, life would be a mistake. The German imagines that even God sings songs."

CODA

THE FINAL JAM WITH DENNIS AND LARRY: A REFLECTIVE IMPROVISATION

LARRY: So, what do you think?

DENNIS: I was amazed at how the combination of our perspectives revealed new ways to look at what music is all about. Learning how much goes on in the brain when making music blew my mind. Now, when I practice, I am more aware and appreciative of my brain's involvement in orchestrating my body movements and the consequential music. It's a bit of a relief to know why learning new skills poses such a challenge. I'm also amazed at how much I take for granted what my eyes are accomplishing when reading music and then how my ears and emotions are being affected. I appreciate how the intellectual and analytical elements are folded into music learning.

Another highlight is being better able to share with others my increased understanding of why stage fright is so normal. I have more confidence to help others bring their authentic and prepared self to public performance. It's clearer how caring about and communicating with audiences can bring inner freedom; getting our minds off ourselves and instead onto our desire to communicate is an act of love that holds higher value. And what the book revealed about musical performance is applicable to public speaking or performance anxiety in general.

What did you think?

LARRY: People often ask me if doing all of this research on music and the brain has influenced the way I listen to or play music. And I think it has. But what I really have come to appreciate from writing this book is how much we still don't know. While we certainly have not explored all the available literature, I feel that we have addressed most of the big questions that help us understand what our brains do during each musical process. But I have more questions:

- What sparks the musical ideas that lead to new compositions or moments of improvisation?
- Are these processes different from other creative processes like writing fiction or painting a picture?
- Why can some people create new music without any apparent effort while others struggle?
- Why can some people just hear a piece of music and recreate it while others must spend hours or days with sheet music, practicing until they feel it is right?
- How is it that some people can compose entirely in their heads while others need to sit down with an instrument or a computer?

We have learned a lot about the circuitry and chemistry of the brain over the past few decades, and that has given us a remarkable framework to explore questions like these.

DENNIS: Learning about the brain's extensive components using the correct labels has helped me feel that understanding the circuitry and chemistry of music is within reach. And I better appreciate how the brain guides and controls us at all hours and then in particular ways when we are involved with music.

I had another "aha" moment when I saw how we may thrive on a variety of experiences that bring us meaning, and how what we really crave is learning. It can be an uphill effort with difficult, hard work, but the sense of accomplishment and subsequent joy, once we "get it," surely leads to a dopamine rush.

LARRY: No doubt. When motivated to practice and play well and achieve a particular goal, the achievement is our reward. For some, that reward is enough to keep practicing either that particular piece of music or more and more music throughout a lifetime. As we discussed, that combination of passion and perseverance must be reinforced by rewards involving dopamine. Those rewards reinforce the desire to learn when we are in the right environment with the right strategy, either developed on our own or with help from a teacher.

DENNIS: Actually, when thinking about the traits of effective music learners, I was shocked to learn that no specific place in the brain processes "passion" and "perseverance." At first I was disappointed, thinking of my parents and teachers training me since my youth to primarily rely on self-motivation and self-discipline by inwardly being strong and commanding myself to achieve goals. But that's not how most people learn. Whether consciously or subconsciously, we humans find strategies that ultimately bring satisfaction through learning.

Whatever the strategy, if it gets a person to practice, study, or move through various exercises in the gym, so be it. Some move through gardening, some through hiking. Some musicians have a "place" to practice, and some have a time of day or night. Having a strategy for practicing with goals and purpose holds strong value for getting it done regularly. And that strategy makes the whole experience of learning pleasurable. This is one reason it is never too late to learn, for no one is willing to put off any substantial happiness.

LARRY: Exactly. And as we mentioned in our discussion about the effects of music practice on processes like neurogenesis, synaptogenesis, and myelination, there is growing evidence that activities like practicing music can drive this plasticity in our brains even in old age. As we said, even old dogs can learn new tricks. It just takes longer!

DENNIS: Another fact I now realize I had taken for granted is that music learning involves the integration of information from our eyes, ears, emotions, thinking, and body movement, and these different types of information can impact one another. Take the example of the alphabet song, where singing letters helps us with memory for those letters.

LARRY: I think the fact that music involves such a high degree of sensory, motor, and cognitive integration is one reason it also has such a powerful effect on our emotions and memories. We are engaging multiple networks within our brains that other activities rarely influence, all at the same time.

DENNIS: Coming back to the question of creativity, I keep thinking about how composing and improvising are linked to different patterns of activation and deactivation in the brain. Realizing that when I'm composing music I'm thinking intensely, yet when I'm improvising I'm attempting to think not at all, makes all the sense in the world. That contrast had not been so clear to me, but now that I grasp the difference, it guides me when to think and when to let go, and how to concentrate in different situations depending on the chosen musical task.

When I'm practicing in order to improvise, I think ardently while developing patterns that I can draw upon when improvising. Then, in my spontaneous state, I can let go, trusting my thought-intensive practicing. On the other hand, when I'm composing, it's mostly deliberate thinking.

LARRY: I think that this is another area where the neuroscience has just scratched the surface. The studies that examine these questions are somewhat hard to compare because the subjects had different backgrounds, the number of subjects in many of the studies wasn't very big, and the study designs were not consistent. Nonetheless, the findings are compelling, and advances in brain imaging will likely lead to even more insights about how we compose and improvise, to what degree these processes are the same or different from person to person, and how we can learn from these differences to enhance the creative process.

Dennis: Another surprise was the organic way the manuscript expanded when we wrote about practicing: one movement grew to three, each worthwhile in its own way. I believe practicing is the central thrust of the book, and in the process of discussing it I realized just how much deliberate practicing challenges the brain.

Larry: I agree! Although listening to music has its own benefits, the process of practicing is the greatest challenge music poses to our brains. As I hope we demonstrated, practicing music is one of the most challenging things a human brain can do. Think of the motor, sensory, and cognitive activity of people who play pipe organs with multiple keyboards with all the various knobs (manuals and stops) and pedals and who sing at the same time. *That* is an incredible accomplishment for any brain!

Dennis: As an organist, I can tell you it isn't easy!

If practicing is process, performance is product. But, as is evident from the answers to our surveys, performance can be influenced by many different variables. Although performers project an energy to an audience, the performers are simultaneously affected by the audience—as if in a conversation. Performance is also affected by whether it's live or for a recording. I know for myself that these are separate artforms. I listen to the room when performing live. I listen to the headphones when recording and then to the speakers on playback to see what the recorded sound was, not necessarily what I thought I sounded like as I was performing.

Larry: As we discussed, what our brains do during practice leads to long-lasting structural and chemical changes in the brain, but they are only the basis for how a performance can go. This is yet another fascinating area of neuroscience, still in its infancy: how brains influence other brains in social settings, in groups of people.

DENNIS: That reminds me of how, when I was young, I listened to music by myself by the hour—and yet I could not wait to share my experiences with others. It was so exciting to share: "I must play you this piece. . . . Listen to this. . . . Do you hear what's happening right there?" As I grew, my expectations and preferences changed, and I still enjoy listening to music, but sharing my joy of music with others is even more pleasurable. To this day music listening, whether by recordings or live, is both an aesthetic and an analytical experience. It can change my perspective, teaching me something new and making the whole experience fun. I think it's another key to effective learning: if we like what we learn, we want to learn more!

LARRY: Yes! Remember how mu-opioid receptors are involved in the liking process? It turns out they play a crucial role in learning and memory.[1] This comes back, I believe, to our discussion on the traits of successful learners and the role of motivation in learning. The same liking-and-wanting cycle that regulates how we experience music also is important for learning, and that can impact how we enjoy the music that we listen to or perform.

I'm struck by the amount of computational power it takes to perceive and respond to music and the fact that we still have much to learn about the underlying circuitry involved in how sound is processed in the auditory cortex before traveling to other areas of the brain. The fact that we almost instantly assign emotional value to simple chords, and the fact that some people have a specific anhedonia for music, support the notion of direct connections between the auditory cortex and the limbic system. This connection highlights the importance of how we react to sound in our daily lives, an ability that must give us certain advantages to survive as well as to use music to pass on emotionally important stories. For the Greeks of Homer's era, the *Iliad* and the *Odyssey* were the basis for the social codes of how to live in ancient Greek society. Assigning greater meaning to the chanted words by playing accompanying melodies in major or minor scales and with different tempos, in my mind, must have helped these works gain their status as important codes to live by.

DENNIS: The success of those ancient, sung stories—and of any music—has to do with finding just the right patterns of notes, chords, rhythm, and all the other parts of music that people find pleasurable. If you sing any tune and change it in only *one* way, the tune is greatly affected. Change the volume, change the harmony, change the rhythm, put it against varied orchestrations, or change the tone quality by the singer or instrumentalist, and the tune is altered, much the way a single chord is altered from major to minor by changing one note. Our levels of enjoyment can be substantially influenced by these minute alterations. So not only do we have specific songs or genres that we like, but we also tend to prefer certain performers playing the same music and even specific recordings. It appears our brains can be picky about what brings us pleasure.

LARRY: That's another great observation! And we still don't understand those little differences, though I think that the powerful connections between music and the mu-opioid and dopamine systems suggest that we have just begun to examine how music therapy can be used. The overlapping signals between pathways in the brain that lead to addiction and the pathways affected by music should be studied further. Music therapy has demonstrated remarkable efficacy for rehabilitation from brain injury. But with greater understanding of how music influences the brain, we may soon see new approaches to using music for addiction and other neuropsychiatric disorders. It's an exciting time for musical neuroscientists. Behold the power of music!

ACKNOWLEDGMENTS

THIS BOOK evolved in part from Larry's talks about music and the brain. Larry was inspired to create these talks by Bobby Heagerty, the former director of community education and outreach for the Oregon Health & Science University Brain Institute, as well as a consummate public educator. The talks evolved in large part due to the support of Amanda Thomas, who has spent years bringing science to public venues through her work with the Oregon Museum of Science and Industry and later as the organizer and emcee of *Science on Tap* and the podcast *A Scientist Walks into a Bar*.

We owe a heartfelt thanks to Jenefer Angell of Passionfruit Projects for urging us to write this book and for her outstanding guidance and editing.

Miranda Martin at Columbia University Press has been extremely helpful, guiding us through the publication process and across the finish line. We could not have asked for a better editor.

We also thank the many people who have offered guidance throughout the process of writing this book. Lou Foltz, a master teacher with vast knowledge of pedagogical matters, contributed substantial feedback. Howard Whitaker, with decades of teaching at a college music conservatory, lent expertise in musical knowledge and exceptional guidance. Jeff Sweeney helped us to understand the best ways to reach our audience, asked the hard questions, and gave invaluable advice. Musician, stage presenter, professor, and writer Kelly Ballard helped us clarify difficult concepts. Harold Gray, a consummate pianist, became our go-to for accurate descriptions of technique, helping us express to readers the physical gestures required to practice and perform without getting too complicated. Colleen Adent, another concert pianist, also confirmed how to write meaningfully for the widest readership. Her additional cheerleading along the way was appreciated deeply. Opera and folk singer and guitarist Peggie Schwarz provided great insights into the process of performing and how to write about it. Brea Murakami, director of the Music Therapy Program at Pacific University, gave us excellent advice about writing for students. We also thank writer and teacher Diane Nichol for her constant support and suggestions, which greatly improved the clarity of our writing.

We feel very lucky to have worked with Susi Davis. She is a remarkable artist, and her illustrations are both informative and fun.

We thank Alanna Sherman for her work in coding the responses to our main survey and for helping us understand what our respondents were telling us.

Finally, we thank all of the respondents to our surveys for helping us understand how people think about music and how their brains work.

APPENDIX A

FIRST SURVEY

The following survey was sent to over one hundred composers, professional and amateur musicians, music teachers, music students, and music lovers. We introduced the idea of the book to them, and they were aware that their responses could be quoted in the book. We examined the responses and highlighted ones that captured ideas expressed by the majority of respondents. Some respondents only answered some of the questions. Answers were binned into categories and then quantified. The x-axes on the following graphs represent the numbers of respondents that gave each category of answer.

A. The origins of music

Q1: Why music?

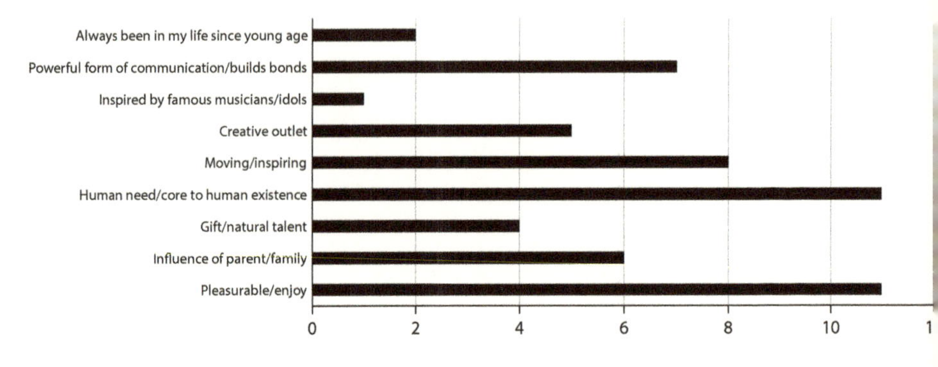

Q2: What motivates you to be involved in making music?

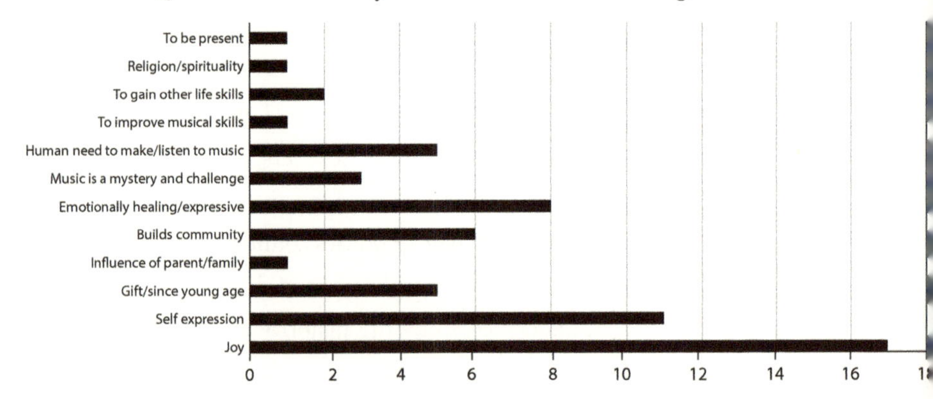

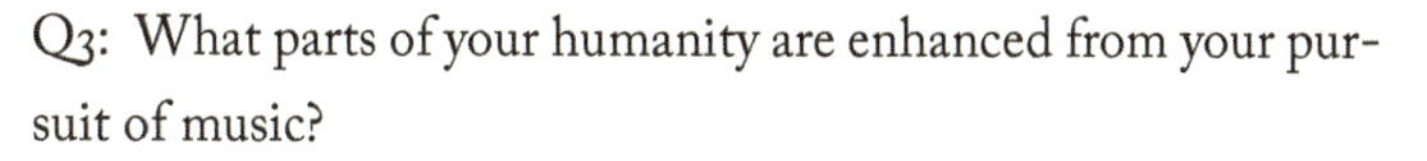

Q3: What parts of your humanity are enhanced from your pursuit of music?

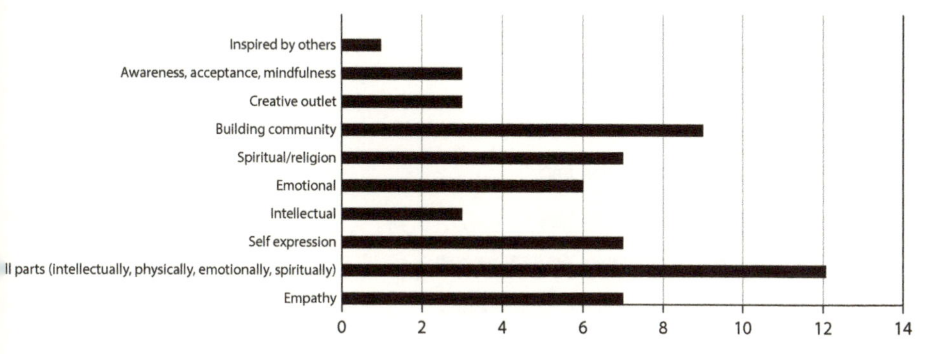

Q4: What role does music play in your human expression and creativity, given that the artists of this world are often the game-changers and rule-breakers?

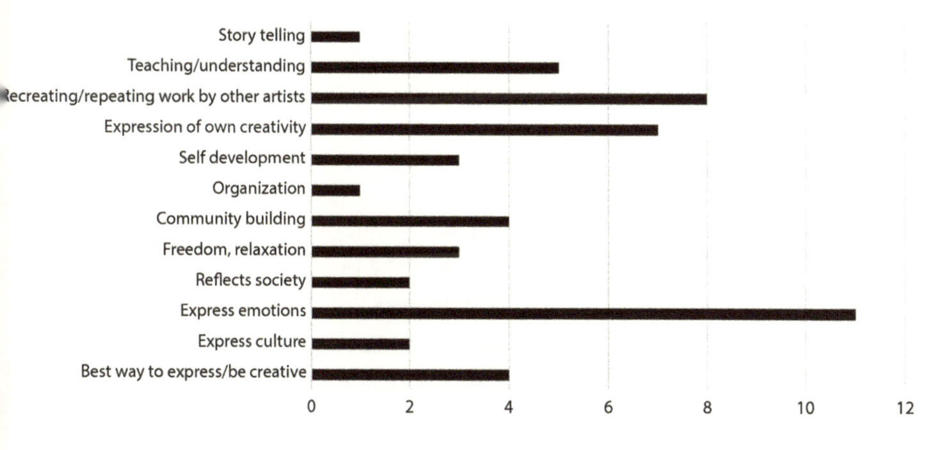

B. How music follows human history

Q1: From your involvement in music, how do you find that it correlates with culture?

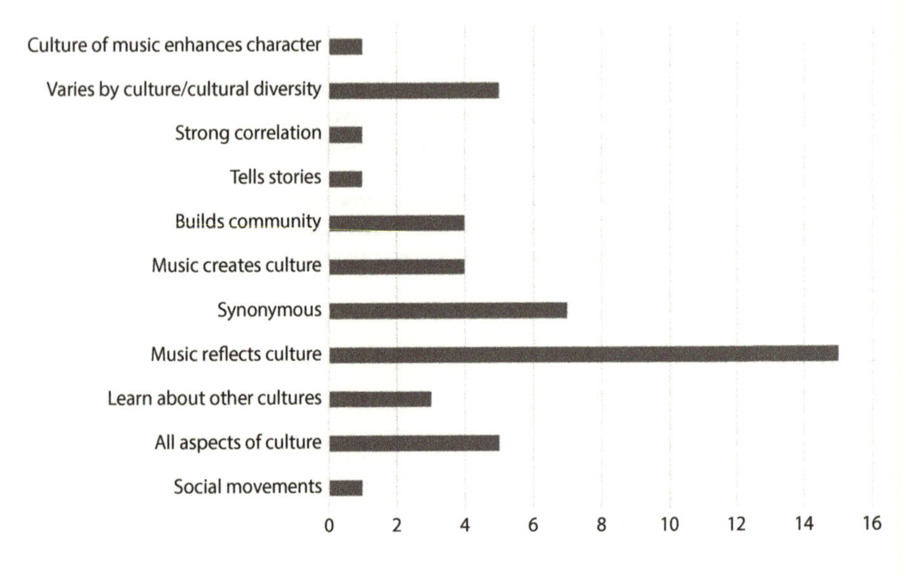

Q2: How is it that music tends to be present at major historical events?

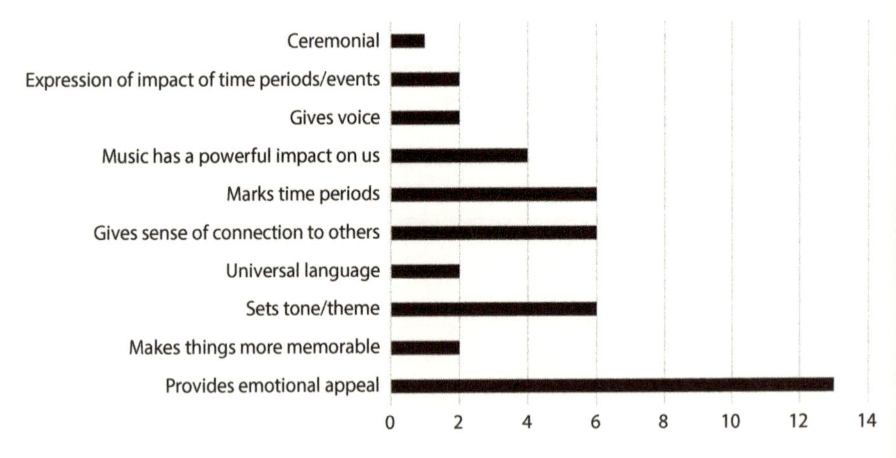

C. How the brain composes

Q1: When do you tend to compose? How do you go about it? What brings inspiration? What motivates you to create?

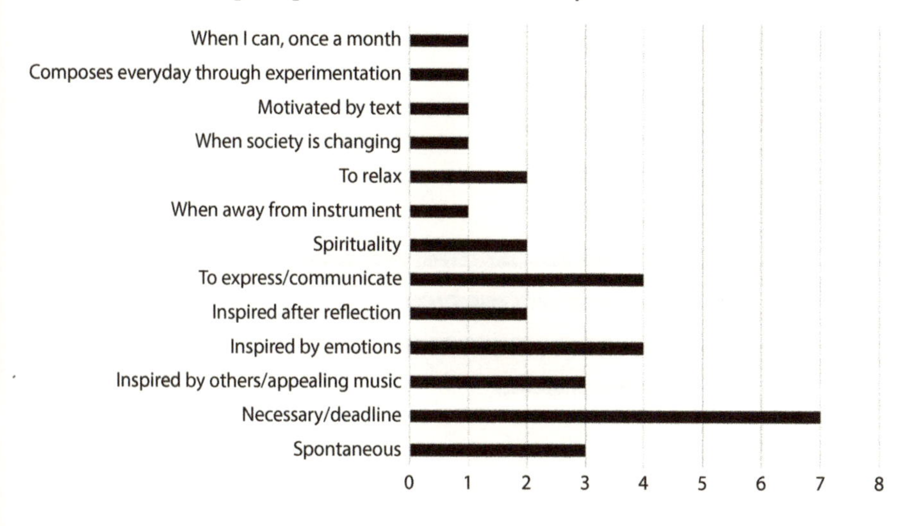

Q2: What are your processes and in what order—text, melody, harmony, rhythm?

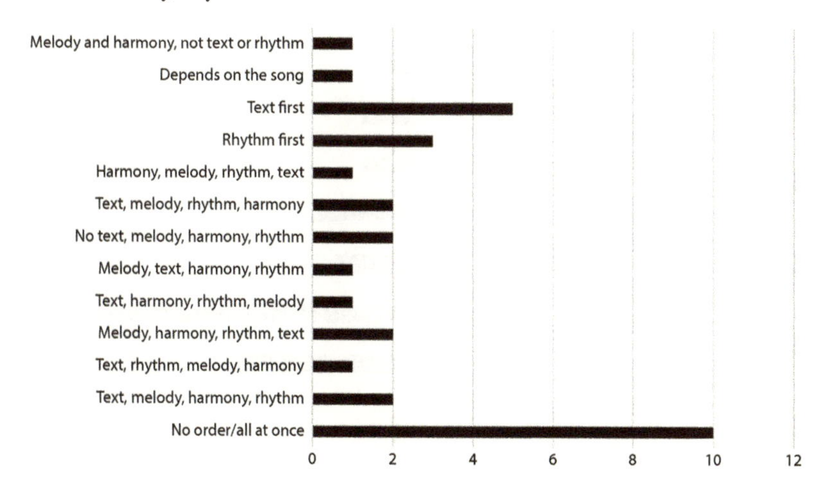

Q3: What other factors, philosophical and/or economic, influence your composing?

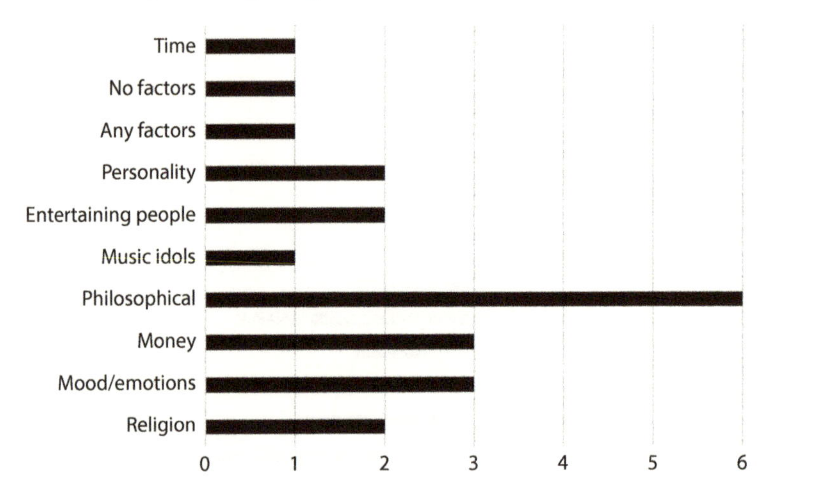

Q4: When creating music, how do vulnerability and wondering how it will be received play into your consciousness?

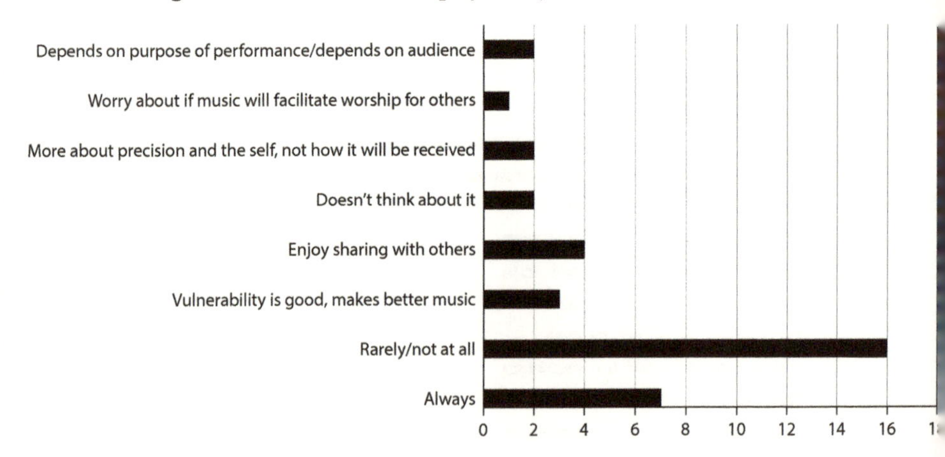

D. How the brain practices

Q1: If you are self-taught, how do you define "practicing"? How do you go about practicing? How do you determine what and when to practice?

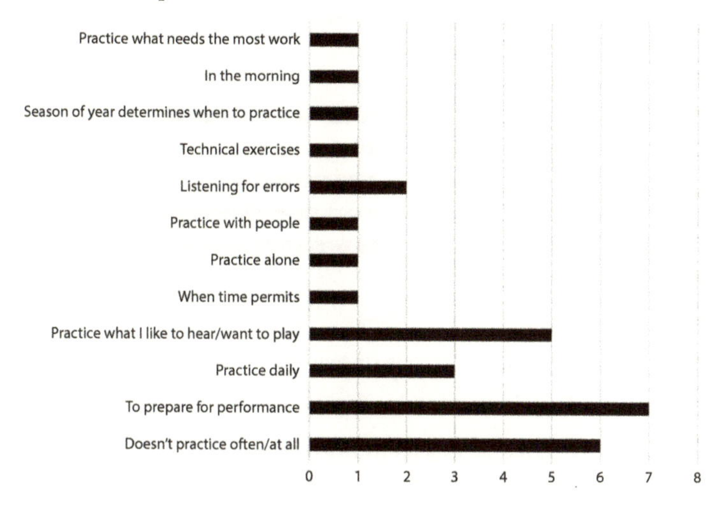

Q2: When you are taking lessons, describe your practice routine. How intentional are you as to when and how to practice?

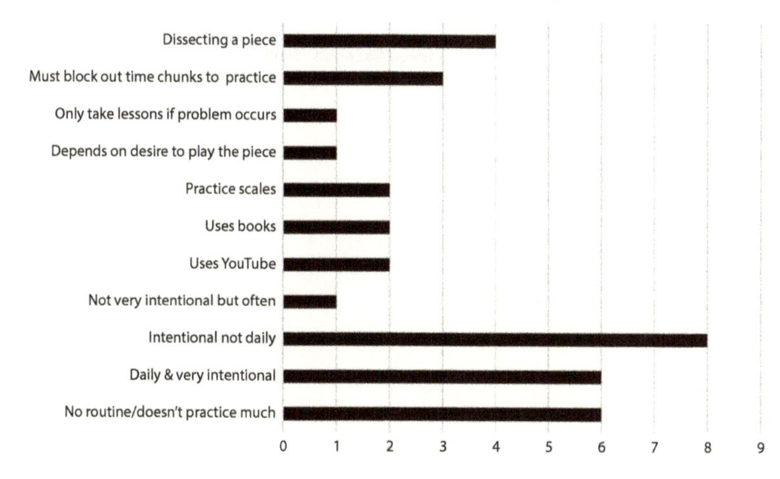

Q3: Describe what you believe is the difference between practicing music and playing music.

Q4: Whether taking lessons or teaching yourself, what are categories and goals for development—sight-reading, playing by ear, various styles of music, improvisation, interpretation, performance piece(s), technique, scales, chords, rhythm studies, chord charts, theory?

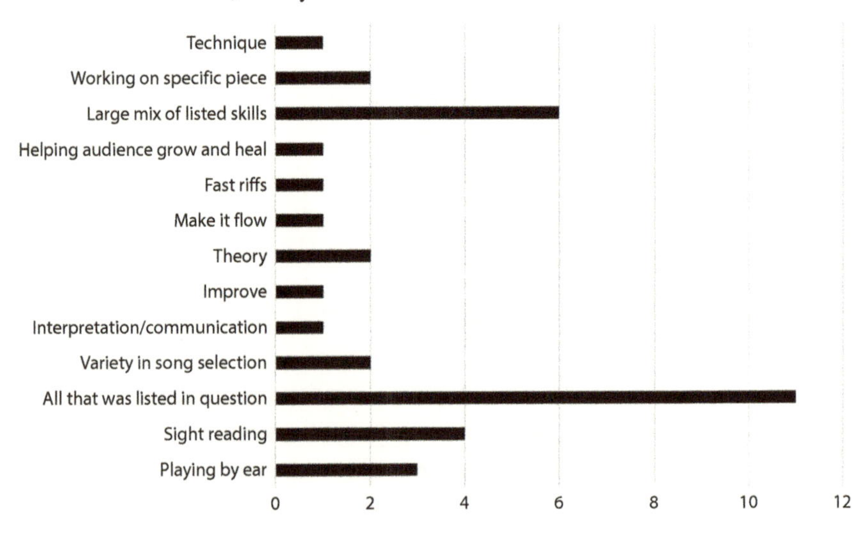

Q5: What do you believe is going on in your brain when you are involved in deliberate practice?

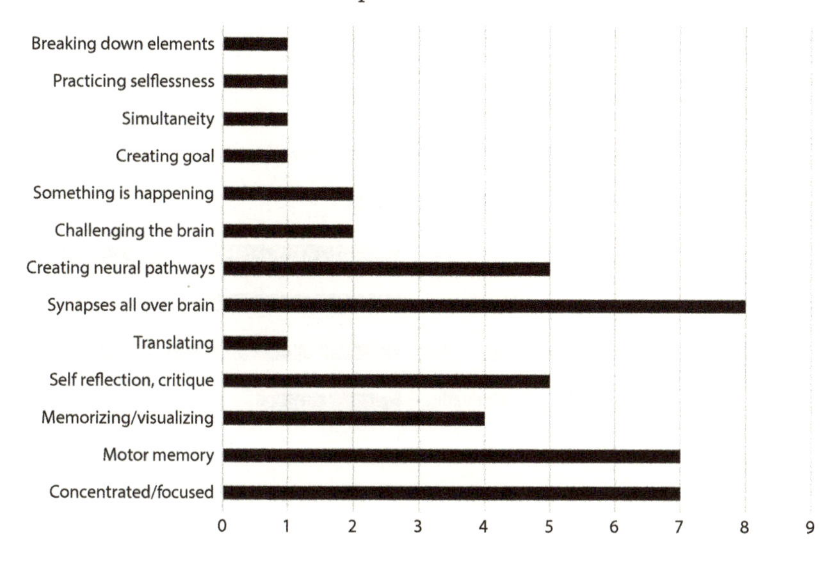

Q6: If possible, conjecture what you believe is going on in your brain when you are performing what has been learned.

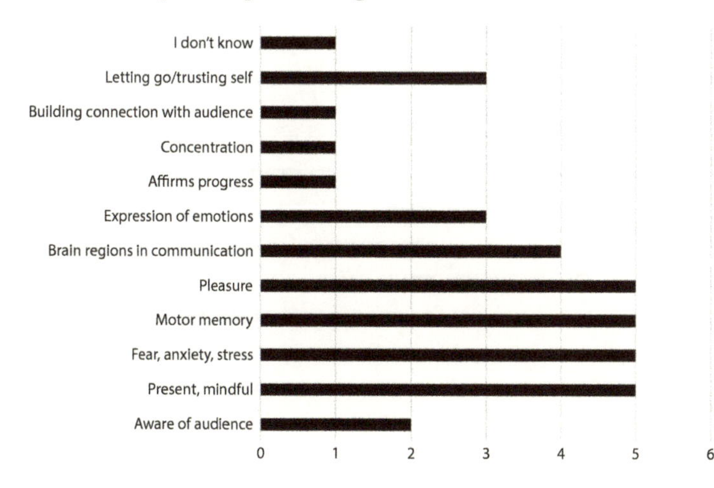

Q7: How do you see the role of failure as a part of practicing?

E. How the brain performs in public—vocal and/or instrumental

Q1: In what mental and emotional space do you find yourself when sharing what you know with others?

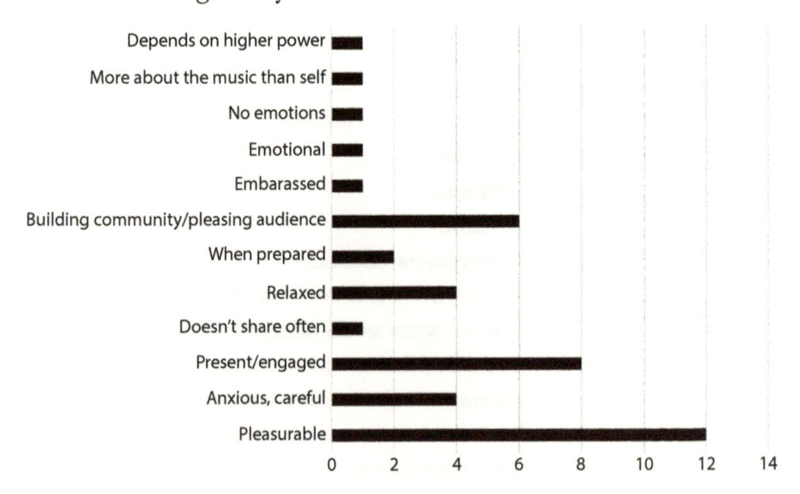

Q2: If you are a vocalist, you are your instrument. This makes music making that much more personal; therefore, when given feedback on your "voice," how can your brain separate critical comments about your instrument from that of your personhood?

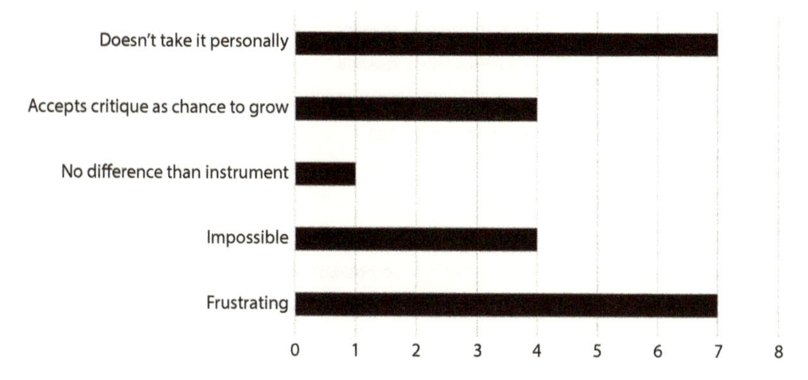

Q3: If you are an instrumentalist, describe how you become one with your instrument and what role you feel you play in achieving the sounds you expect.

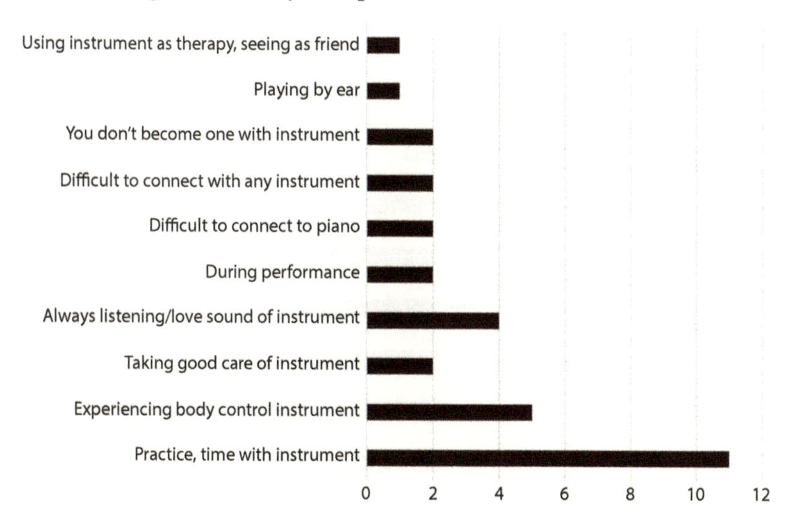

Q4: How does stage fright play into performing for others? Have you fear, joy, concentration, concern about the listeners, worry over extramusical factors, or complete relaxation and enjoyment? How do you manage your brain to be your friend?

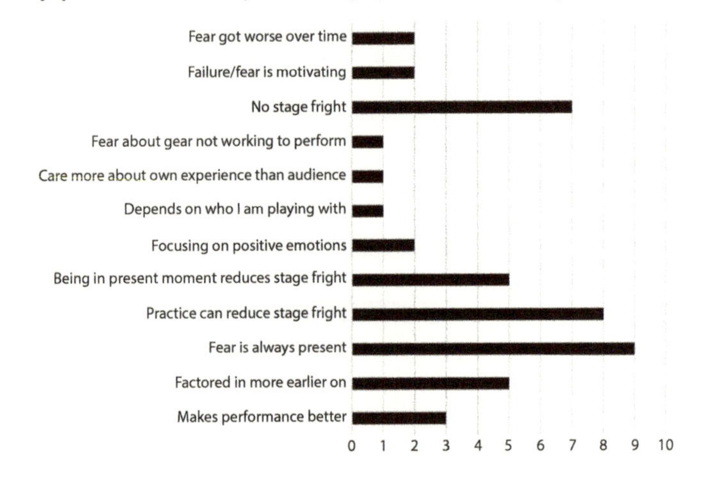

Q5: If you are in a recording session, how is your attitude different from when in a live audience situation? How is your brain operating when comparing and contrasting recording for posterity and for a live performance?

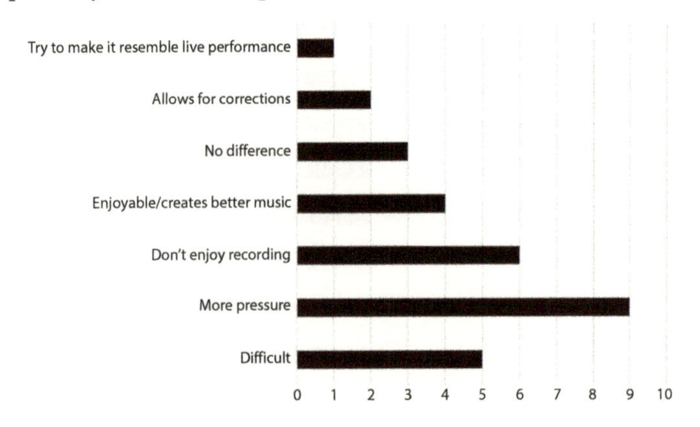

Q6: If you perform with others, what goes through your mind when you are hearing what members of your group are playing? Do you consciously alter your playing to adjust to theirs?

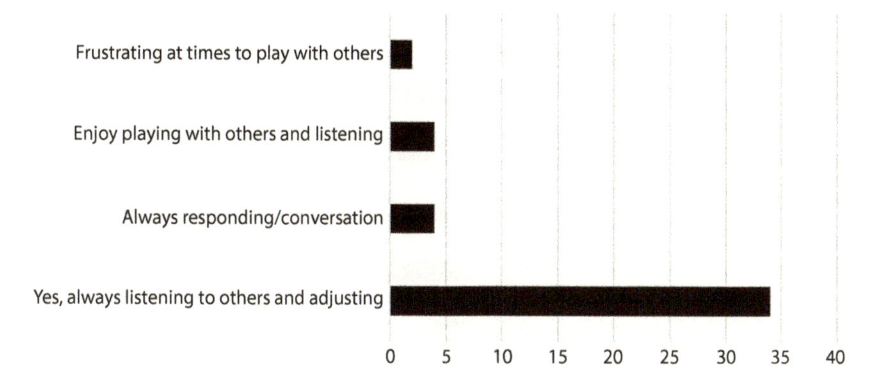

Q7: How does your emotional response to the music affect your playing?

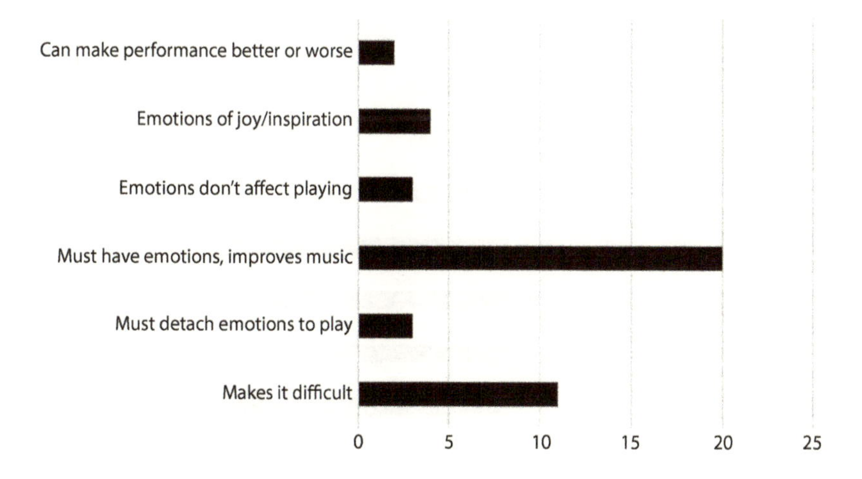

F: How the brain performs in the conducting mode

Q1: When conducting a rehearsal with components to manage simultaneously, what is going on in your brain?

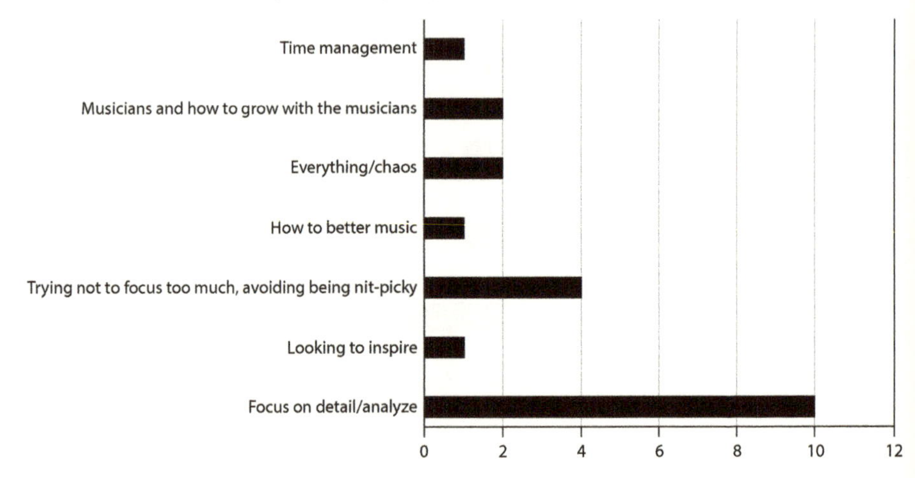

Q2: When conducting a public performance, what is going on in your brain?

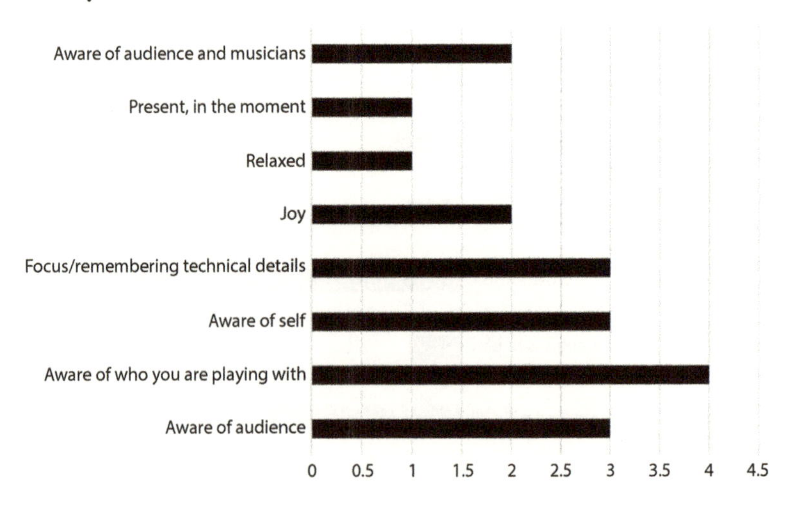

G. How the brain listens to music

Q1: What do you figure is the difference in your brain between hearing music as background and listening consciously and with purpose? How do you know when you are doing one over the other?

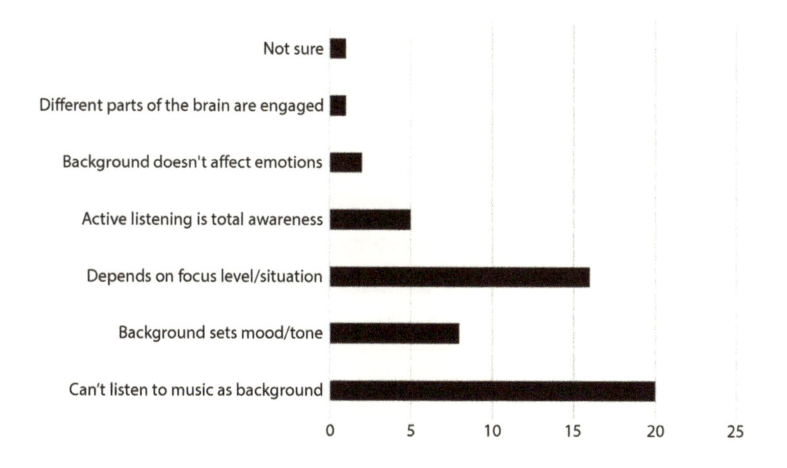

Q2: What is your relationship to music when experiencing it via recordings (vinyl, cassette, cd, and streaming)?

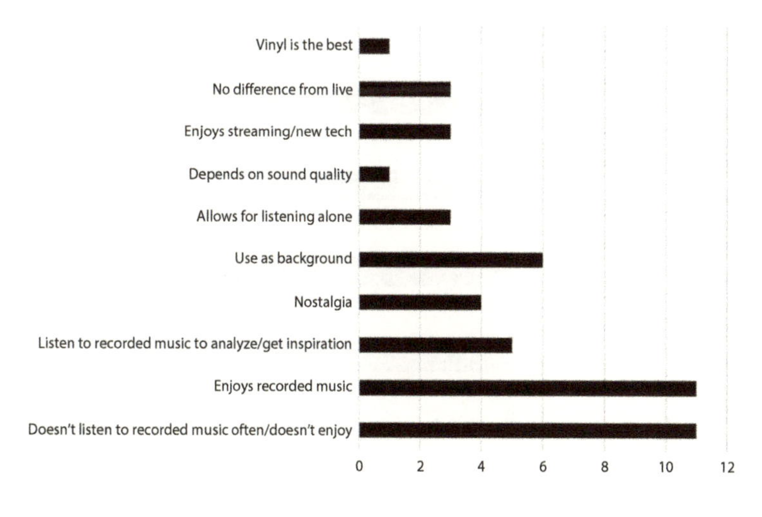

Q3: What is your relationship to music when experiencing it live?

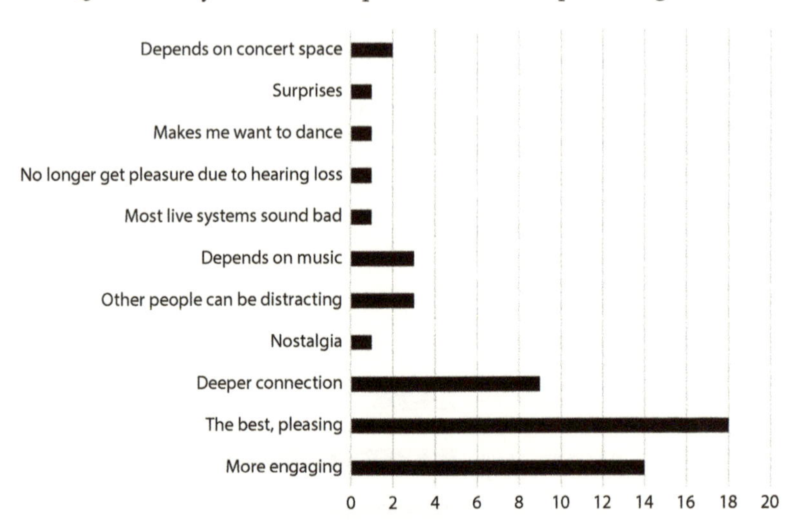

Q4: When listening to scored and rehearsed music, like classical, how does the technical quality of the performance affect you, especially if you are acquainted with the composer's intention?

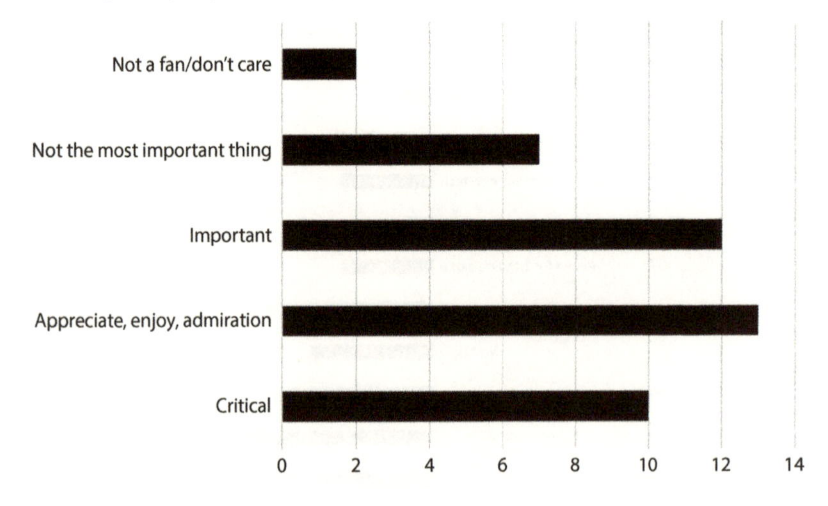

Q5: When listening to improvisational music, like jazz and blues, what grabs your ear significantly—performer passion (soul), technical skill, and/or what else?

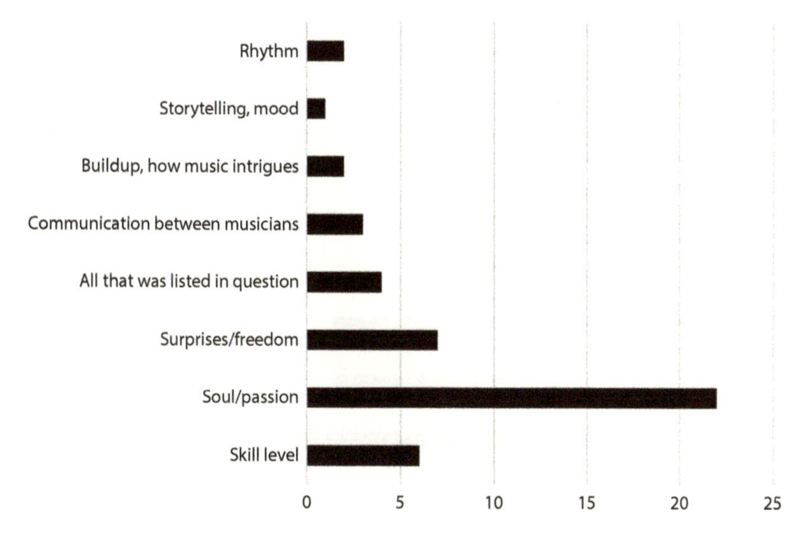

Q6: How do you feel that the social climate (those around you) affects your listening to music, whether recorded or live?

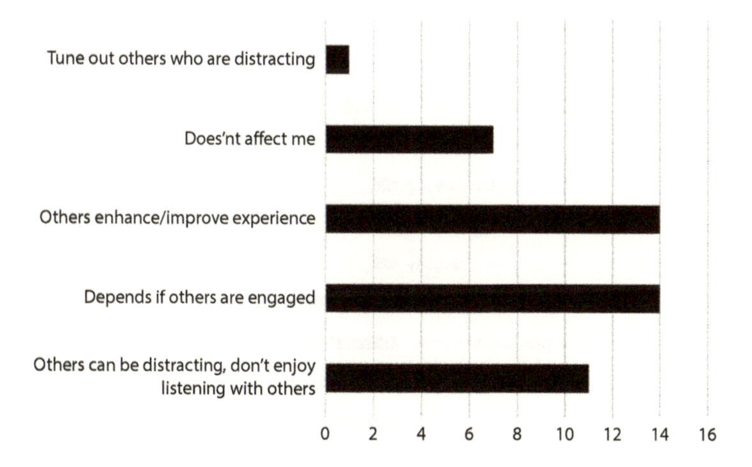

H. The benefits of music to human health and cognition

Q1: When involved in music—whether composing, practicing, performing, or listening—what do you surmise are benefits to your health and cognition?

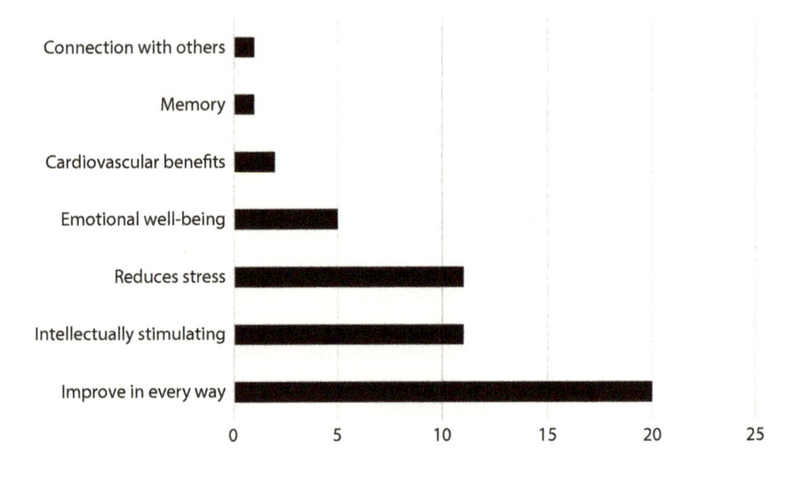

Q2: What effect does music involvement have in your life? In other words, how do musical pursuits transfer to other aspects of your life?

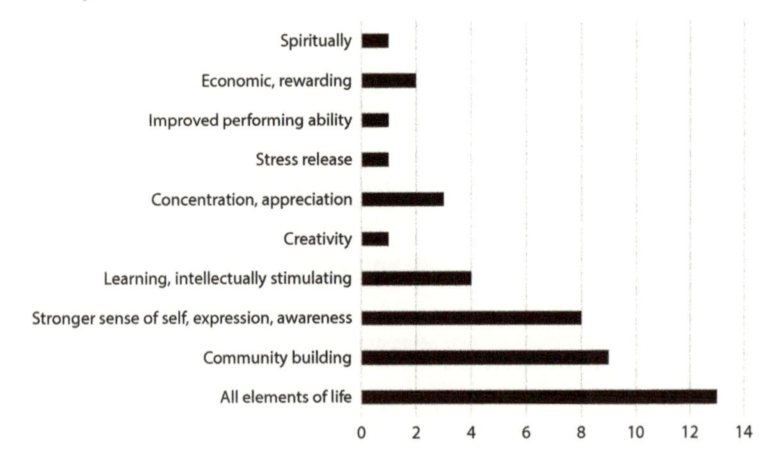

Q3: Describe any social factors that come to mind through the engagement with music.

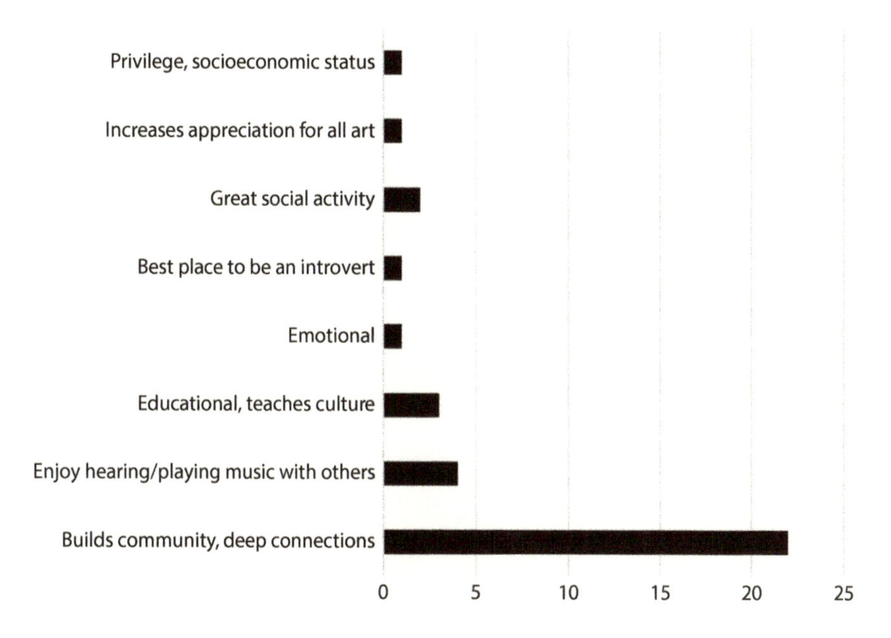

APPENDIX B

SECOND SURVEY

The following survey was sent to people who responded to the first survey (appendix A). They were aware that their responses could be quoted in the book. As with the first survey, we examined the responses and highlighted ones that captured ideas expressed by the majority of respondents.

- Realizing how involved the brain is when taking in music, please share one or more stories regarding the emotional impact that certain music can have on you.
- Reflect on times long ago and/or recently when a musical selection grabbed you, and describe what that was like for you. Also, please describe the circumstances of that experience.

- If you are a performer or conductor, feel free to share about when you are involved in that certain piece that always grabs you—how it affects your intellect, your heart, even your body.

NOTES

1. WHAT IS MUSIC, AND WHY DOES IT EXIST?

1. F. A. Azevedo et al., "Equal Numbers of Neuronal and Nonneuronal Cells Make the Human Brain an Isometrically Scaled-Up Primate Brain," *Journal of Comparative Neurology* 513 (2009): 532–41.
2. Y. Tang et al., "Total Regional and Global Number of Synapses in the Human Brain Neocortex," *Synapse* 41 (2001): 258–73.
3. A. Eklund, T. E. Nichols, and H. Knutsson, "Cluster Failure: Why fMRI Inferences for Spatial Extent Have Inflated False-Positive Rates," *Proceedings of the National Academy of Sciences of the United States of America* 113, no. 28 (2016): 7900–7905.
4. S. Norman-Haignere, N. G. Kanwisher, and J. H. McDermott, "Distinct Cortical Pathways for Music and Speech Revealed by Hypothesis-Free Voxel Decomposition," *Neuron* 88 (2015): 1281–96.
5. C. M. Schlebusch et al., "Southern African Ancient Genomes Estimate Modern Human Divergence to 350,000 to 260,000 Years Ago," *Science* 358 (2017): 652–55.
6. N. Conard, M. Malina, and S. C. Münzel, "New Flutes Document the Earliest Musical Tradition in Southwestern Germany," *Nature* 460 (2009): 737–40; T. Higham et al., "Testing Models for the Beginnings

of the Aurignacian and the Advent of Figurative Art and Music: The Radiocarbon Chronology of Geißenklösterle," *Journal of Human Evolution* 62 (2012): 664–76; I. Morley, *The Prehistory of Music: Human Evolution, Archaeology, and the Origins of Musicality* (Oxford: Oxford University Press, 2013).

7. B. Toškan, "Remains of Large Mammals from Divje Babe I: Stratigraphy, Taxonomy, and Biometry," *Opera Instituti Archaeologici Sloveniae* 13 (2007): 221–78.
8. C. G. Diedrich, "'Neanderthal Bone Flutes': Simply Products of Ice Age Spotted Hyena Scavenging Activities on Cave Bear Cubs in European Cave Bear Dens," *Royal Society Open Science* 2 (2015): 140022.
9. R. Fink, "The Neanderthal Flute and the Origins of the Scale: Fang or Flint? A Response," *Studies in Music Archaeology* 3 (2003): 83–87.
10. B. Nettl, "An Ethnomusicologist Contemplates Universals in Musical Sound and Musical Culture," in *The Origins of Music*, ed. N. L. Wallin, B. Merker, and S. Brown (Cambridge, MA: MIT Press, 2000), 468.
11. A. Dell'Anna, M. Leman, and A. Berti, "Musical Interaction Reveals Music as Embodied Language," *Frontiers in Neuroscience* 15 (2021): 667838.
12. C. Darwin, *On the Origin of Species* (London: John Murray, 1859).
13. J. Schulkin and G. B. Raglan, "The Evolution of Music and Human Social Capability," *Frontiers in Neuroscience* 8 (2014): 292.
14. L. Bernstein, *The Unanswered Question: Six Talks at Harvard by Leonard Bernstein*, rev. ed. (Cambridge, MA: Harvard University Press, 1981).
15. S. Mithen, *The Singing Neanderthals: The Origins of Music, Language, Mind and Body* (London: Weidenfeld and Nicolson, 2005), 197.
16. I. McGilchrist, *The Master and His Emissary: The Divided Brain and the Making of the Western World* (New Haven, CT: Yale University Press, 2009), 103.
17. G. O'Neill, "Humming, Whistling, Singing, and Yelling in Piraha Context and Channels of Communication in FDG," *Pragmatics* 24 (2014): 349–75.
18. From *New Jubilee Songs as Sung by the Fisk Jubilee Singers* by Frederick J. Work and John Wesley Work Jr. (1901). Composer unknown.
19. C. A. Green Jr., *OurStory: Putting Color Back into His-Story: What We Dragged Out of Slavery with Us* (West Conshohocken, PA: Infinity, 2006).

2. HOW YOUR BRAIN COMPOSES MUSIC

1. G. Loewenstein, "The Psychology of Curiosity: A Review and Reinterpretation," *Psychological Bulletin* 116 (1994): 75–98; P. Y. Oudeyer and F. Kaplan, "What Is Intrinsic Motivation? A Typology of Computational Approaches," *Frontiers in Neurorobotics* 1 (2007): 6.
2. C. Kidd and B. Y. Hayden, "The Psychology and Neuroscience of Curiosity," *Neuron* 88 (2015): 449–60.
3. M. E. Gross, C. M. Zedelius, and J. W. Schooler, "Cultivating an Understanding of Curiosity as a Seed for Creativity," *Current Opinion in Behavioral Sciences* 35 (2020): 77–82.
4. R. Martin, *Beethoven's Hair: An Extraordinary Historical Odyssey and a Scientific Mystery Solved* (New York: Crown, 2001).
5. R. M. Krebs et al., "The Novelty Exploration Bonus and Its Attentional Modulation," *Neuropsychologia* 47 (2009): 2272–81; L. Shen, A. Fishbach, and C. K. Hsee, "The Motivating-Uncertainty Effect: Uncertainty Increases Resource Investment in the Process of Reward Pursuit," *Journal of Consumer Research* 41 (2015): 1301–15.
6. T. Daikoku, "Depth and the Uncertainty of Statistical Knowledge on Musical Creativity Fluctuate over a Composer's Lifetime," *Frontiers in Computational Neuroscience* 13 (2019): 27; T. Daikoku, "Time-Course Variation of Statistics Embedded in Music: Corpus Study on Implicit Learning and Knowledge," *PLoS One* 13 (2018): e0196493.
7. S. L. Bengtsson, M. Csíkszentmihályi, and F. Ullén, "Cortical Regions Involved in the Generation of Musical Structures During Improvisation in Pianists," *Journal of Cognitive Neuroscience* 19 (2007): 830–42.
8. C. J. Limb and A. R. Braun, "Neural Substrates of Spontaneous Musical Performance: An FMRI Study of Jazz Improvisation," *PLoS One* 3, no. 2 (2008): e1679.
9. G. F. Donnay et al., "Neural Substrates of Interactive Musical Improvisation: An FMRI Study of 'Trading Fours' in Jazz," *PLoS One* 9 (2014): e88665.
10. A. Belden et al., "Improvising at Rest: Differentiating Jazz and Classical Music Training with Resting State Functional Connectivity," *NeuroImage* 207 (2020): 116384.
11. K. C. Barrett et al., "Classical Creativity: A Functional Magnetic Resonance Imaging (fMRI) Investigation of Pianist and Improviser Gabriela Montero," *NeuroImage* 209 (2020): 116496.

12. K. Dhakal et al., "Higher Node Activity with Less Functional Connectivity During Musical Improvisation," *Brain Connectivity* 9 (2019): 296–309.
13. M. Hickey, "Can Improvisation Be 'Taught'? A Call for Free Improvisation in Our Schools," *International Journal of Music Education* 27 (2009): 285.
14. Quoted in D. Borgo, *Sync or Swarm: Improvising Music in a Complex Age* (New York: Bloomsbury Academic, 2005), 9.
15. C. Arkin et al., "Gray Matter Correlates of Creativity in Musical Improvisation," *Frontiers in Human Neuroscience* 13 (2019): 169.
16. E. Ozdemir, A. Norton, and G. Schlaug, "Shared and Distinct Neural Correlates of Singing and Speaking," *NeuroImage* 33 (2006): 628–35.
17. S. B. Eickhoff et al., "Anatomical and Functional Connectivity of Cytoarchitectonic Areas Within the Human Parietal Operculum," *Journal of Neuroscience* 30 (2010): 6409–21.
18. D. J. Levitin and S. T. Grafton, "Measuring the Representational Space of Music with fMRI: A Case Study with Sting," *Neurocase* 22 (2016): 548–57.
19. J. Lu et al., "The Brain Functional State of Music Creation: An fMRI Study of Composers," *Scientific Reports* 5 (2015): 12277.
20. J. Lu et al., "The Multiple-Demand System in the Novelty of Musical Improvisation: Evidence from an MRI Study on Composers," *Frontiers in Neuroscience* 11 (2017): 695.
21. A. Woolgar et al., "The Multiple-Demand System but Not the Language System Supports Fluid Intelligence," *Nature Human Behaviour* 2 (2018): 200–204.

3. PRACTICING MUSIC, PART I: THE PARTNERSHIP OF MOTIVATED MUSIC STUDENTS AND MOTIVATED MUSIC TEACHERS

1. K. M. Bieszczad, and N. M. Weinberger, "Learning Strategy Trumps Motivational Level in Determining Learning-Induced Auditory Cortical Plasticity," *Neurobiology of Learning and Memory* 93 (2010): 229–39.
2. L. Pauwels, S. Chalavi, and S. P. Swinnen, "Aging and Brain Plasticity," *Aging* 10 (2018): 1789–90.
3. A. L. Duckworth et al., "Grit: Perseverance and Passion for Long-Term Goals," *Journal of Personality and Social Psychology* 92 (2007): 1087–101.

4. K. R. Von Culin, E. Tsukayama, and A. L. Duckworth, "Unpacking Grit: Motivational Correlates of Perseverance and Passion for Long-Term Goals," *Journal of Positive Psychology* 9 (2014): 306–12.
5. K. Rimfeld et al., "True Grit and Genetics: Predicting Academic Achievement from Personality," *Journal of Personality and Social Psychology* 111 (2016): 780–89.
6. E. M. Tucker-Drob et al., "Genetically-Mediated Associations Between Measures of Childhood Character and Academic Achievement," *Journal of Personality and Social Psychology* 111 (2016): 790–815.
7. T. G. Reio Jr. and A. Wiswell, "Field Investigation of the Relationship Among Adult Curiosity, Workplace Learning, and Job Performance," *Human Resource Development Quarterly* 11 (2000): 5–30; M. J. Kang et al., "The Wick in the Candle of Learning: Epistemic Curiosity Activates Reward Circuitry and Enhances Memory," *Psychological Science* 20 (2009): 963–73; M. J. Gruber, B. D. Gelman, and C. Ranganath, "States of Curiosity Modulate Hippocampus-Dependent Learning via the Dopaminergic Circuit," *Neuron* 84 (2014): 486–96; A. E. Stahl and L. Feigenson, "Cognitive Development: Observing the Unexpected Enhances Infants' Learning and Exploration," *Science* 348 (2015): 91–94.
8. S. Wade and C. Kidd, "The Role of Prior Knowledge and Curiosity in Learning," *Psychonomic Bulletin and Review* 26 (2019): 1377–87.
9. L. Moccia et al., "The Experience of Pleasure: A Perspective Between Neuroscience and Psychoanalysis," *Frontiers in Human Neuroscience* 12 (2018): 359.
10. P. Sörqvist and J. E. Marsh, "How Concentration Shields Against Distraction," *Current Directions in Psychological Science* 24 (2015): 267–72.

4. PRACTICING MUSIC, PART II: UNDERSTANDING THE NEUROSCIENCE

1. F. Bouhali et al., "Reading Music and Words: The Anatomical Connectivity of Musicians' Visual Cortex," *NeuroImage* 212 (2020): 116666.
2. P. H. Rudebeck et al., "Specialized Representations of Value in the Orbital and Ventrolateral Prefrontal Cortex: Desirability Versus Availability of Outcomes," *Neuron* 95 (2017): 1208–20.e5.
3. S. Zhang, P. Liu, and T. Feng, "To Do It Now or Later: The Cognitive Mechanisms and Neural Substrates Underlying Procrastination," *Wiley Interdisciplinary Reviews Cognitive Science* 10 (2019): e1492.

4. P. Ball, *The Music Instinct: How Music Works and Why We Can't Do Without It* (Oxford: Oxford University Press, 2010), 228.
5. The notion of "muscle memory" is often applied to all types of practice situations, not just musical, but it is a misnomer. Muscles have no memory: a muscle that is not connected to the motor or somatosensory cortex already described cannot perform any skilled actions, nor does a muscle store skilled routines. So, when people say "muscle memory," they are really describing the routine itself, the coordinated collection of activities involved in each element of music that requires memory, which influences the control of those muscles.
6. M. Walker, *Why We Sleep: Unlocking the Power of Sleep and Dreams* (New York: Scribner, 2017); B. Rasch and J. Born, "About Sleep's Role in Memory," *Physiological Reviews* 93 (2013): 681–766.

5. PRACTICING MUSIC, PART III: CHANGING YOUR BRAIN TO GET IT RIGHT

1. J. Altman, "Are New Neurons Formed in the Brains of Adult Mammals?," *Science* 135 (1962): 1127–28.
2. S. Owji and M. M. Shoja, "The History of Discovery of Adult Neurogenesis," *Clinical Anatomy* 33 (2020): 41–55.
3. C. L. Pytte et al., "Adult Neurogenesis Is Associated with the Maintenance of a Stereotyped, Learned Motor Behavior," *Journal of Neuroscience* 32 (2012): 7052–57.
4. G. Kempermann, "Adult Neurogenesis: An Evolutionary Perspective," *Cold Spring Harbor Perspectives in Biology* 8 (2015): a018986.
5. L. Vaquero et al., "Structural Neuroplasticity in Expert Pianists Depends on the Age of Musical Training Onset," *NeuroImage* 126 (2016): 106–19.
6. M. Groussard et al., "The Effects of Musical Practice on Structural Plasticity: The Dynamics of Grey Matter Changes," *Brain and Cognition* 90 (2014): 174–80.
7. M. S. Oechslin et al., "Hippocampal Volume Predicts Fluid Intelligence in Musically Trained People," *Hippocampus* 23 (2013): 552–58.
8. M. B. Kennedy, "Synaptic Signaling in Learning and Memory," *Cold Spring Harbor Perspectives in Biology* 8 (2013): a016824.
9. C. F. Heaney and J. W. Kinney, "Role of GABAB Receptors in Learning and Memory and Neurological Disorders," *Neuroscience and Biobehavioral Reviews* 63 (2016): 1–28.

10. K. Rosenkranz, A. Williamon, and J. C. Rothwell, "Motorcortical Excitability and Synaptic Plasticity Is Enhanced in Professional Musicians," *Journal of Neuroscience* 27 (2007): 5200–5206.
11. B. Zalc, D. Goujet, and D. Colman, "The Origin of the Myelination Program in Vertebrates," *Current Biology* 18 (2008): R511–12.
12. G. A. Craig, S. Yoo, T. Y. Du, and J. Xiao, "Plasticity in Oligodendrocyte Lineage Progression: An OPC Puzzle on Our Nerves," *European Journal of Neuroscience* 54 (2021): 5747–61; D. K. Dansu, S. Sauma, and P. Casaccia, "Oligodendrocyte Progenitors as Environmental Biosensors," *Seminars in Cell and Developmental Biology* 116 (2021): 38–44.
13. G. Schlaug et al., "Increased Corpus Callosum Size in Musicians," *Neuropsychologia* 33 (1995): 1047–55.
14. A. H. Oztürk et al., "Morphometric Comparison of the Human Corpus Callosum in Professional Musicians and Non-musicians by Using in Vivo Magnetic Resonance Imaging," *Journal of Neuroradiology* 29 (2002): 29–34; D. J. Lee, Y. Chen, and G. Schlaug, "Corpus Callosum: Musician and Gender Effects," *NeuroReport* 14 (2003): 205–9; K. L. Hyde et al., "Musical Training Shapes Structural Brain Development," *Journal of Neuroscience* 29 (2009): 3019–25.
15. S. L. Bengtsson et al., "Extensive Piano Practicing Has Regionally Specific Effects on White Matter Development," *Nature Neuroscience* 8 (2005): 1148–50.
16. E. Moore et al., "Diffusion Tensor MRI Tractography Reveals Increased Fractional Anisotropy (FA) in Arcuate Fasciculus Following Music-Cued Motor Training," *Brain and Cognition* 116 (2017): 40–46.
17. V. B. Penhune, "Musical Expertise and Brain Structure: The Causes and Consequences of Training," in *The Oxford Handbook of Music and the Brain*, ed. M. H. Thaut and D. A. Hodges (Oxford: Oxford University Press, 2019), 419–38.
18. P. Ragert et al., "Superior Tactile Performance and Learning in Professional Pianists: Evidence for Meta-Plasticity in Musicians," *European Journal of Neuroscience* 19 (2004): 473–78.
19. M. Hund-Georgiadis and D. Y. von Cramon, "Motor-Learning-Related Changes in Piano Players and Non-musicians Revealed by Functional Magnetic-Resonance Signals," *Experimental Brain Research* 125 (1999): 417–25; D. Watanabe, T. Savion-Lemieux, and V. B. Penhune, "The Effect of Early Musical Training on Adult Motor Performance: Evidence for a Sensitive Period in Motor Learning," *Experimental Brain Research* 176 (2007): 332–40.

20. S. Swaminathan and E. G. Schellenberg, "Music Training and Cognitive Abilities: Associations, Causes, and Consequences," in *The Oxford Handbook of Music and the Brain*, 645–70.

6. HOW YOUR BRAIN PERFORMS MUSIC

1. R. J. Zatorre, J. L. Chen, and V. B. Penhune, "When the Brain Plays Music: Auditory-Motor Interactions in Music Perception and Production," *Nature Reviews Neuroscience* 8 (2007): 547–58.
2. R. Rojiani et al., "Communication of Emotion via Drumming: Dual-Brain Imaging with Functional Near-Infrared Spectroscopy," *Social Cognitive and Affective Neuroscience* 13 (2018): 1047–57.
3. R. K. Wolf, "Embodiment and Ambivalence: Emotion in South Asian Muharram Drumming," *Yearbook for Traditional Music* 32 (2000): 81–116.
4. Y. Hou et al., "The Averaged Inter-brain Coherence Between the Audience and a Violinist Predicts the Popularity of Violin Performance," *NeuroImage* 211 (2020): 116655.
5. V. Gallese et al., "Action Recognition in the Premotor Cortex," *Brain* 119 (1996): 593–609; G. Rizzolatti and L. Fogassi, "The Mirror Mechanism: Recent Findings and Perspectives," *Philosophical Transactions of the Royal Society B Biological Sciences* 369 (2014): 20130420.
6. V. Gallese, "The Manifold Nature of Interpersonal Relations: The Quest for a Common Mechanism," *Philosophical Transactions of the Royal Society B Biological Sciences* 358 (2003): 517–28.
7. I. Molnar-Szakacs and K. Overy, "Music and Mirror Neurons: From Motion to 'E'motion," *Social Cognitive and Affective Neuroscience* 1 (2006): 235–41.
8. D. Swarbrick et al., "How Live Music Moves Us: Head Movement Differences in Audiences to Live Versus Recorded Music," *Frontiers in Psychology* 9 (2019): 2682.
9. A. Stiller, "Toward a Biology of Music," *OPUS* 35 (1987): 12.
10. D. Weinstein et al., "Group Music Performance Causes Elevated Pain Thresholds and Social Bonding in Small and Large Groups of Singers," *Evolution and Human Behavior* 37 (2016): 152–58.
11. R. I. Dunbar et al., "Performance of Music Elevates Pain Threshold and Positive Affect: Implications for the Evolutionary Function of Music," *Evolutionary Psychology* 10 (2012): 688–702.

12. J. Martín-Fernández et al., "Music Style Not Only Modulates the Auditory Cortex, but Also Motor Related Areas," *Neuroscience* 457 (2021): 88–102.
13. M. Nusseck, M. Zander, and C. Spahn, "Music Performance Anxiety in Young Musicians: Comparison of Playing Classical or Popular Music," *Medical Problems of Performing Artists* 30 (2015): 30–37.

7. HOW YOUR BRAIN LISTENS TO MUSIC

1. C. M. Wessinger et al., "Tonotopy in Human Auditory Cortex Examined with Functional Magnetic Resonance Imaging," *Human Brain Mapping* 5 (1997): 18–25.
2. S. Koelsch, "Toward a Neural Basis of Music Perception—A Review and Updated Model," *Frontiers in Psychology* 2 (2011): 110.
3. M. E. Klein and R. J. Zatorre, "A Role for the Right Superior Temporal Sulcus in Categorical Perception of Musical Chords," *Neuropsychologia* 49 (2011): 878–87.
4. R. G. Crowder, "Perception of the Major/Minor Distinction: I. Historical and Theoretical Foundations," *Psychomusicology: A Journal of Research in Music Cognition* 4 (1984): 3–12.
5. M. Suzuki et al., "Discrete Cortical Regions Associated with the Musical Beauty of Major and Minor Chords," *Cognitive, Affective, and Behavioral Neuroscience* 8 (2008): 126–31.
6. A. Kolchinsky et al., "The Minor Fall, the Major Lift: Inferring Emotional Valence of Musical Chords Through Lyrics," *Royal Society Open Science* 4 (2017): 170952.
7. M. P. Kastner and R. G. Crowder, "Perception of the Major/Minor Distinction: IV. Emotional Connotations in Young Children," *Music Perception* 8 (1990): 189–201; P. G. Hunter and E. G. Schellenberg, "Music and Emotion," *Music Perception* 36 (2010): 129–64; R. G. Crowder, J. S. Reznick, and S. L. Rosenkrantz, "Perception of the Major/Minor Distinction: V. Preferences Among Infants," *Bulletin of the Psychonomic Society* 29 (1991): 187–88.
8. T. Phillips and A. D'Angour, *Music, Text, and Culture in Ancient Greece* (Oxford: Oxford University Press, 2018).
9. E. J. Allen et al., "Representations of Pitch and Timbre Variation in Human Auditory Cortex," *Journal of Neuroscience* 37 (2017): 1284–93.

10. S. Samson, "Neuropsychological Studies of Musical Timbre," *Annals of the New York Academy of Sciences* 999 (2003): 144–51.
11. T. B. Janzen and M. H. Thaut, "Cerebral Organization of Music Processing," in *The Oxford Handbook of Music and the Brain*, ed. M. H. Thaut and D. A. Hodges (Oxford: Oxford University Press, 2019), 89–121.
12. M. Molinari et al., "Sensorimotor Transduction of Time Information Is Preserved in Subjects with Cerebellar Damage," *Brain Research Bulletin* 67 (2005): 448–58.
13. S. Paquette et al., "The Cerebellum's Contribution to Beat Interval Discrimination," *NeuroImage* 163 (2017): 177–82.
14. S. Nozaradan et al., "Specific Contributions of Basal Ganglia and Cerebellum to the Neural Tracking of Rhythm," *Cortex* 95 (2017): 156–68.
15. S. E. Trehub, "Musical Predispositions in Infancy," *Annals of the New York Academy of Sciences* 930 (2001): 1–16; L. J. Trainor, "Are There Critical Periods for Musical Development?," *Developmental Psychobiology* 46 (2005): 262–78.
16. R. J. Zatorre and V. N. Salimpoor, "From Perception to Pleasure: Music and Its Neural Substrates," *Proceedings of the National Academy of Sciences of the United States of America* 110 (2013): 10430–37.
17. J. C. Fernández-Miranda et al., "Asymmetry, Connectivity, and Segmentation of the Arcuate Fascicle in the Human Brain," *Brain Structure and Function* 220 (2015): 1665–80.
18. F. Giovannelli et al., "The Effect of Music on Corticospinal Excitability Is Related to the Perceived Emotion: A Transcranial Magnetic Stimulation Study," *Cortex* 49 (2013): 702–10; K. Michaelis, M. Wiener, and J. C. Thompson, "Passive Listening to Preferred Motor Tempo Modulates Corticospinal Excitability," *Frontiers in Human Neuroscience* 8 (2014): 252; J. Stupacher et al., "Musical Groove Modulates Motor Cortex Excitability: A TMS Investigation," *Brain and Cognition* 82 (2013): 127–36.
19. I. Cross, "Musicality and the Human Capacity for Culture," *Musicae Scientiae* 12 (2008): 147–67; N. L. Wallin, B. Merker, and S. Brown, eds., *The Origins of Music* (Cambridge, MA: MIT Press, 2000).
20. S. Koelsch et al., "The Roles of Superficial Amygdala and Auditory Cortex in Music-Evoked Fear and Joy," *NeuroImage* 81 (2013): 49–60; M. Lehne, M. Rohrmeier, and S. Koelsch, "Tension-Related Activity in the Orbitofrontal Cortex and Amygdala: An fMRI Study with Music," *Social Cognitive and Affective Neuroscience* 9 (2014): 1515–23; S. Koelsch

and S. Skouras. "Functional Centrality of Amygdala, Striatum and Hypothalamus in a 'Small-World' Network Underlying Joy: An fMRI Study with Music," *Human Brain Mapping* 35 (2014): 3485–98.

21. S. Koelsch, "Brain Correlates of Music-Evoked Emotions," *Nature Reviews Neuroscience* 15 (2014): 170–80.
22. A. J. Blood and R. J. Zatorre, "Intensely Pleasurable Responses to Music Correlate with Activity in Brain Regions Implicated in Reward and Emotion," *Proceedings of the National Academy of Sciences of the United States of America* 98 (2001): 11818–23; V. N. Salimpoor et al., "Anatomically Distinct Dopamine Release During Anticipation and Experience of Peak Emotion to Music," *Nature Neuroscience* 14 (2011): 257–62; V. N. Salimpoor et al., "Interactions Between the Nucleus Accumbens and Auditory Cortices Predict Music Reward Value," *Science* 340 (2013): 216–19.
23. K. J. Pallesen et al., "Emotion Processing of Major, Minor, and Dissonant Chords: A Functional Magnetic Resonance Imaging Study," *Annals of the New York Academy of Sciences* 1060 (2005): 450–53; S. Koelsch et al., "Investigating Emotion with Music: An fMRI Study," *Human Brain Mapping* 27 (2006): 239–50; M. T. Mitterschiffthaler et al., "A Functional MRI Study of Happy and Sad Affective States Induced by Classical Music," *Human Brain Mapping* 28 (2007): 1150–62; E. Brattico et al., "A Functional MRI Study of Happy and Sad Emotions in Music with and Without Lyrics," *Frontiers in Psychology* 2 (2011): 308.
24. V. Putkinen et al., "Decoding Music-Evoked Emotions in the Auditory and Motor Cortex," *Cerebral Cortex* 31 (2021): 2549–60.
25. L. Jäncke, "Music, Memory and Emotion," *Journal of Biology* 7 (2008): 21.
26. R. Harris and B. M. de Jong, "Cerebral Activations Related to Audition-Driven Performance Imagery in Professional Musicians," *PLoS One* 9 (2014): e93681.
27. T. Hasegawa et al., "Learned Audio-Visual Cross-Modal Associations in Observed Piano Playing Activate the Left Planum Temporale: An fMRI Study," *Cognitive Brain Research* 20 (2004): 510–18; S. Baumann et al., "A Network for Audio-Motor Coordination in Skilled Pianists and Non-musicians," *Brain Research* 1161 (2007): 65–78; R. Bianco et al., "Neural Networks for Harmonic Structure in Music Perception and Action," *NeuroImage* 142 (2016): 454–64.
28. A. Wuttke-Linnemann et al., "Sex-Specific Effects of Music Listening on Couples' Stress in Everyday Life," *Scientific Reports* 9 (2019): 4880.

7. HOW YOUR BRAIN LISTENS TO MUSIC

29. S. Moreno et al., "Short-Term Second Language and Music Training Induces Lasting Functional Brain Changes in Early Childhood," *Child Development* 86 (2015): 394–406.
30. F. H. Rauscher, G. L. Shaw, and K. N. Ky, "Music and Spatial Task Performance," *Nature* 365 (1993): 611.
31. J. Pietschnig, M. Voracek, and A. K. Formann, "Mozart Effect—Shmozart Effect: A Meta-analysis," *Intelligence* 38 (2010): 314–23.
32. W. F. Thompson, E. G. Schellenberg, and G. Husain, "Arousal, Mood, and the Mozart Effect," *Psychological Science* 12 (2001): 248–51.
33. Fernández-Miranda et al., "Asymmetry, Connectivity, and Segmentation."

8. WHY YOUR BRAIN LIKES MUSIC

1. L. Dubé and J. Le Bel, "The Content and Structure of Laypeople's Concept of Pleasure," *Cognition and Emotion* 17 (2003): 263–95.
2. T. Schäfer et al., "The Psychological Functions of Music Listening," *Frontiers in Psychology* 4 (2013): 511.
3. D. Zald and R. J. Zatorre, "Music," in *Neurobiology of Sensation and Reward*, ed. D. Zald, R. J. Zatorre, and J. A. Gottfried (Boca Raton, FL: CRC Press/Taylor and Francis, 2011).
4. A. G. DiFeliceantonio et al., "Enkephalin Surges in Dorsal Neostriatum as a Signal to Eat," *Current Biology* 22 (2012): 1918–24.
5. A. Mallik, M. L. Chanda, and D. J. Levitin, "Anhedonia to Music and Mu-Opioids: Evidence from the Administration of Naltrexone," *Scientific Reports* 7 (2017): 41952.
6. N. L. Stone et al., "An Analysis of Endocannabinoid Concentrations and Mood Following Singing and Exercise in Healthy Volunteers," *Frontiers in Behavioral Neuroscience* 12 (2018): 269.
7. C. T. Tart, "Marijuana Intoxication Common Experiences," *Nature* 226 (1970): 701–4; B. Green, D. Kavanagh, and R. Young, "Being Stoned: A Review of Self-Reported Cannabis Effects," *Drug and Alcohol Review* 22 (2003): 453–60.
8. M. S. Lim et al., "A Cross-Sectional Survey of Young People Attending a Music Festival: Associations Between Drug Use and Musical Preference," *Drug and Alcohol Review* 27 (2008): 439–41; T. Van Havere et al., "Drug Use and Nightlife: More Than Just Dance Music," *Substance Abuse Treatment Prevention and Policy* 6 (2011): 18; J. J. Palamar,

M. Griffin-Tomas, and D. C. Ompad, "Illicit Drug Use Among Rave Attendees in a Nationally Representative Sample of U.S. High School Seniors," *Drug and Alcohol Dependence* 152 (2015): 24–31.

9. K. C. Berridge and M. L. Kringelbach, "Pleasure Systems in the Brain," *Neuron* 86 (2015): 646–64.
10. V. N. Salimpoor et al., "Anatomically Distinct Dopamine Release During Anticipation and Experience of Peak Motion to Music," *Nature Neuroscience* 14 (2011): 257–62.
11. L. Ferreri et al., "Dopamine Modulates the Reward Experiences Elicited by Music," *Proceedings of the National Academy of Sciences of the United States of America* 116 (2019): 3793–98.
12. A. J. Sihvonen et al., "Neural Architectures of Music—Insights from Acquired Amusia," *Neuroscience and Biobehavioral Reviews* 107 (2019): 104–14.
13. E. Mas-Herrero et al., "The Impact of Visual Art and Emotional Sounds in Specific Musical Anhedonia," *Progress in Brain Research* 237 (2018): 399–413.
14. A. M. Belfi and P. Loui, "Musical Anhedonia and Rewards of Music Listening: Current Advances and a Proposed Model," *Annals of the New York Academy of Sciences* 1464 (2020): 99–114.
15. N. Martínez-Molina et al., "Neural Correlates of Specific Musical Anhedonia," *Proceedings of the National Academy of Sciences of the United States of America* 113 (2016): E7337–E45.
16. P. Loui et al., "White Matter Correlates of Musical Anhedonia: Implications for Evolution of Music," *Frontiers in Psychology* 8 (2017): 1664.
17. C. Ahrends, "Does Excessive Music Practicing Have Addiction Potential?," *Psychomusicology: Music, Mind, and Brain* 27 (2017): 191–202; N. Schmuziger et al., "Is There Addiction to Loud Music? Findings in a Group of Non-professional Pop/Rock Musicians," *Audiology Research* 2 (2012): e11.
18. K. C. Berridge and T. E. Robinson, "Liking, Wanting, and the Incentive-Sensitization Theory of Addiction," *American Psychologist* 71 (2016): 670–79.
19. M. Reybrouck, P. Podlipniak, and D. Welch, "Music Listening as Coping Behavior: From Reactive Response to Sense-Making," *Behavioral Sciences* 10 (2020): 119.
20. P. N. Juslin and P. Laukka, "Expression, Perception, and Induction of Musical Emotions: A Review and a Questionnaire Study of Everyday Listening," *Journal of New Music Research* 33 (2004): 217–38.

21. M. E. Sachs, A. Damasio, and A. Habibi, "The Pleasures of Sad Music: A Systematic Review," *Frontiers in Human Neuroscience* 9 (2015): 404.
22. J. Levinson, *Music, Art and Metaphysics: Essays in Philosophical Aesthetics* (Oxford: Oxford University Press, 1990).
23. S. Garrido and E. Schubert, "Individual Differences in the Enjoyment of Negative Emotion in Music: A Literature Review and Experiment," *Music Perception* 28 (2011): 279–96; A. J. M. Van den Tol and J. Edwards, "Exploring a Rationale for Choosing to Listen to Sad Music When Feeling Sad," *Psychology of Music* 41 (2013): 440–65; L. Taruffi and S. Koelsch, "The Paradox of Music-Evoked Sadness: An Online Survey," *PLoS One* 9 (2014): e110490.
24. D. Huron and J. K. Vuoskoski, "On the Enjoyment of Sad Music: Pleasurable Compassion Theory and the Role of Trait Empathy," *Frontiers in Psychology* 11 (2020): 1060.
25. W. T. Harbaugh, U. Mayr, and D. R. Burghart, "Neural Responses to Taxation and Voluntary Giving Reveal Motives for Charitable Donations," *Science* 316 (2007): 1622–25.
26. R. J. Harris et al., "Young Men's and Women's Different Autobiographical Memories of the Experience of Seeing Frightening Movies on a Date," *Media Psychology* 2 (2000): 245–68; T. Eerola, J. K. Vuoskoski, and H. Kautiainen, "Being Moved by Unfamiliar Sad Music Is Associated with High Empathy," *Frontiers in Psychology* 7 (2016): 1176; J. K. Vuoskoski and T. Eerola, "Can Sad Music Really Make You Sad? Indirect Measures of Affective States Induced by Music and Autobiographical Memories," *Psychology of Aesthetics, Creativity, and the Arts* 6 (2012): 204.
27. P. J. Rentfrow and S. D. Gosling. "The Do Re Mi's of Everyday Life: The Structure and Personality Correlates of Music Preferences," *Journal of Personality and Social Psychology* 84 (2003): 1236–56; A. C. North, "Individual Differences in Musical Taste," *American Journal of Psychology* 123 (2010): 199–208; D. Boer et al., "How Shared Preferences in Music Create Bonds Between People: Values as the Missing Link," *Personality and Social Psychology Bulletin* 37 (2011): 1159–71; A. Bonneville-Roussy et al., "Music Through the Ages: Trends in Musical Engagement and Preferences from Adolescence Through Middle Adulthood," *Journal of Personality and Social Psychology* 105 (2013): 703–17.
28. P. J. Rentfrow and J. A. McDonald, "Music Preferences and Personality," in Handbook of Music and Emotion, ed. P. N. Juslin and J. Sloboda (Oxford: Oxford University Press, 2010), 669–95.

29. D. M. Greenberg et al., "Musical Preferences Are Linked to Cognitive Styles," *PLoS One* 10 (2015): e0131151.
30. J. K. Vuoskoski and T. Eerola, "The Role of Mood and Personality in the Perception of Emotions Represented by Music," *Cortex* 47 (2011): 1099–106; J. K. Vuoskoski et al., "Who Enjoys Listening to Sad Music and Why?," Music Perception 29 (2012): 311–17; A. C. Miu and F. R. Balteş, "Empathy Manipulation Impacts Music-Induced Emotions: A Psychophysiological Study on Opera," *PloS One* 7 (2012): e30618; T. C. Rabinowitch, I. Cross, and P. Burnard, "Long-Term Musical Group Interaction Has a Positive Influence on Empathy in Children," *Psychology of Music* 41 (2013): 484–98; H. Egermann and S. McAdams, "Empathy and Emotional Contagion as a Link Between Recognized and Felt Emotions in Music Listening," *Music Perception: An Interdisciplinary Journal* 31 (2013): 139–56.
31. J. H. McDermott et al., "Indifference to Dissonance in Native Amazonians Reveals Cultural Variation in Music Perception," *Nature* 535 (2016): 547–50.
32. M. J. McPherson et al., "Perceptual Fusion of Musical Notes by Native Amazonians Suggests Universal Representations of Musical Intervals," *Nature Communications* 11 (2020): 2786.
33. B. Kaneshiro et al., "Characterizing Listener Engagement with Popular Songs Using Large-Scale Music Discovery Data," *Frontiers in Psychology* 8 (2017): 416.
34. D. Omigie and J. Ricci, "Accounting for Expressions of Curiosity and Enjoyment During Music Listening," *Psychology of Aesthetics, Creativity, and the Arts*, advance online publication, 2022.

CODA

1. W. J. Meilandt, E. Barea-Rodriguez, S. A. Harvey, J. L. Martinez Jr, "Role of Hippocampal CA3 Mu-Opioid Receptors in Spatial Learning and Memory," *Journal of Neuroscience* 24 (2004): 2953–62.

BIBLIOGRAPHY

FIRST MOVEMENT

Azevedo, F. A., L. R. Carvalho, L. T. Grinberg, J. M. Farfel, R. E. Ferretti, R. R. Leite, W. Jacob Filho, R. Lent, and S. Herculano-Houzel. "Equal Numbers of Neuronal and Nonneuronal Cells Make the Human Brain an Isometrically Scaled-Up Primate Brain." *Journal of Comparative Neurology* 513 (2009): 532–41.

Bernstein, L. *The Unanswered Question: Six Talks at Harvard by Leonard Bernstein*. Rev. ed. Cambridge, MA: Harvard University Press, 1981.

Conard, N., M. Malina, and S. C. Münzel. "New Flutes Document the Earliest Musical Tradition in Southwestern Germany." *Nature* 460 (2009): 737–40.

Darwin, C. *On the Origin of Species*. London: John Murray, 1859.

Dell'Anna, A., M. Leman, and A. Berti. "Musical Interaction Reveals Music as Embodied Language." *Frontiers in Neuroscience* 15 (2021): 667838.

Diedrich, C. G. "'Neanderthal Bone Flutes': Simply Products of Ice Age Spotted Hyena Scavenging Activities on Cave Bear Cubs in European Cave Bear Dens." *Royal Society Open Science* 2 (2015): 140022.

Eklund, A., T. E. Nichols, and H. Knutsson. "Cluster Failure: Why fMRI Inferences for Spatial Extent Have Inflated False-Positive Rates." *Proceedings of the National Academy of Sciences of the United States of America* 113 (2016): 7900–7905.

Fink, R. "The Neanderthal Flute and the Origins of the Scale: Fang or Flint? A Response." *Studies in Music Archaeology* 3 (2003): 83–87.

Green, C. A., Jr. *OurStory: Putting Color Back into His-Story: What We Dragged Out of Slavery with Us.* West Conshohocken, PA: Infinity, 2006.

Higham, T., L. Basell, R. Jacobi, R. Wood, C. B. Ramsey, and N. J. Conard. "Testing Models for the Beginnings of the Aurignacian and the Advent of Figurative Art and Music: The Radiocarbon Chronology of Geißenklösterle." *Journal of Human Evolution* 62 (2012): 664–76.

McGilchrist, I. *The Master and His Emissary: The Divided Brain and the Making of the Western World.* New Haven, CT: Yale University Press, 2009.

Mithen, S. *The Singing Neanderthals: The Origins of Music, Language, Mind, and Body.* London: Weidenfeld and Nicolson, 2005.

Morley, I. *The Prehistory of Music: Human Evolution, Archaeology, and the Origins of Musicality.* Oxford: Oxford University Press, 2013.

Nettl, B. "An Ethnomusicologist Contemplates Universals in Musical Sound and Musical Culture." In *The Origins of Music*, ed. N. L. Wallin, B. Merker, and S. Brown, 463–72. Cambridge, MA: MIT Press, 2000.

Norman-Haignere, S., N. G. Kanwisher, and J. H. McDermott. "Distinct Cortical Pathways for Music and Speech Revealed by Hypothesis-Free Voxel Decomposition." *Neuron* 88 (2015): 1281–96.

O'Neill, G. "Humming, Whistling, Singing, and Yelling in Piraha Context and Channels of Communication in FDG." *Pragmatics* 24 (2014): 349–75.

Schlebusch, C. M., H. Malmström, T. Günther, P. Sjödin, A. Coutinho, H. Edlund, A. R. Munters, M. Vicente, M. Steyn, H. Soodyall, M. Lombard, and M. Jakobsson. "Southern African Ancient Genomes Estimate Modern Human Divergence to 350,000 to 260,000 Years Ago." *Science* 358 (2017): 652–55.

Schulkin, J., and G. B. Raglan. "The Evolution of Music and Human Social Capability." *Frontiers of Neuroscience* 8 (2014): 292.

Tang, Y., J. R. Nyengaard, D. M. De Groot, and H. J. Gundersen. "Total Regional and Global Number of Synapses in the Human Brain Neocortex." *Synapse* 41 (2001): 258–73.

Toškan, B. "Remains of Large Mammals from Divje Babe I: Stratigraphy, Taxonomy, and Biometry." *Opera Instituti Archaeologici Sloveniae* 13 (2007): 221–78.

SECOND MOVEMENT

Arkin, C., E. Przysinda, C. W. Pfeifer, T. Zeng, and P. Loui. "Gray Matter Correlates of Creativity in Musical Improvisation." *Frontiers in Human Neuroscience* 13 (2019): 169.

Barrett, K. C., F. S. Barrett, P. Jiradejvong, S. K. Rankin, A. T. Landau, and C. J. Limb. "Classical Creativity: A Functional Magnetic Resonance Imaging (fMRI) Investigation of Pianist and Improviser Gabriela Montero." *NeuroImage* 209 (2020): 116496.

Belden, A., T. Zeng, E. Przysinda, S. A. Anteraper, S. Whitfield-Gabrieli, and P. Loui. "Improvising at Rest: Differentiating Jazz and Classical Music Training with Resting State Functional Connectivity." *NeuroImage* 207 (2020): 116384.

Bengtsson, S. L., M. Csíkszentmihályi, and F. Ullén. "Cortical Regions Involved in the Generation of Musical Structures During Improvisation in Pianists." *Journal of Cognitive Neuroscience* 19 (2007): 830–42.

Borgo, D. *Sync or Swarm: Improvising Music in a Complex Age*. New York: Bloomsbury Academic, 2005.

Daikoku, T. "Depth and the Uncertainty of Statistical Knowledge on Musical Creativity Fluctuate over a Composer's Lifetime." *Frontiers in Computational Neuroscience* 13 (2019): 27.

Daikoku, T. "Time-Course Variation of Statistics Embedded in Music: Corpus Study on Implicit Learning and Knowledge." *PLoS One* 13 (2018): e0196493.

Dhakal, K., M. Norgaard, B. M. Adhikari, K. S. Yun, and M. Dhamala. "Higher Node Activity with Less Functional Connectivity During Musical Improvisation." *Brain Connectivity* 9 (2019): 296–309.

Donnay, G. F., S. K. Rankin, M. Lopez-Gonzalez, P. Jiradejvong, and C. J. Limb. "Neural Substrates of Interactive Musical Improvisation: An FMRI Study of 'Trading Fours' in Jazz." *PLoS One* 9 (2014): e88665.

Eickhoff, S. B., S. Jbabdi, S. Caspers, A. R. Laird, P. T. Fox, K. Zilles, and T. E. Behrens. "Anatomical and Functional Connectivity of Cytoarchitectonic Areas Within the Human Parietal Operculum." *Journal of Neuroscience* 30 (2010): 6409–21.

Gross, M. E., C. M. Zedelius, and J. W. Schooler. "Cultivating an Understanding of Curiosity as a Seed for Creativity." *Current Opinion in Behavioral Sciences* 35 (2020): 77–82.

Hickey, M. "Can Improvisation Be 'Taught'? A Call for Free Improvisation in Our Schools." *International Journal of Music Education* 27 (2009): 285–99.

Kidd, C., and B. Y. Hayden. "The Psychology and Neuroscience of Curiosity." *Neuron* 88 (2015): 449–60.

Krebs, R. M., B. H. Schott, H. Schütze, and E. Düzel. "The Novelty Exploration Bonus and Its Attentional Modulation." *Neuropsychologia* 47 (2009): 2272–81.

Levitin, D. J., and S. T. Grafton. "Measuring the Representational Space of Music with fMRI: A Case Study with Sting." *Neurocase* 22 (2016): 548–57.

Limb, C. J., and A. R. Braun. "Neural Substrates of Spontaneous Musical Performance: An FMRI Study of Jazz Improvisation." *PLoS One* 27 (2008): 3(2): e1679.

Loewenstein, G. "The Psychology of Curiosity: A Review and Reinterpretation." *Psychological Bulletin* 116 (1994): 75–98.

Lu, J., H. Yang, H. He, S. Jeon, C. Hou, A. C. Evans, and D. Yao. "The Multiple-Demand System in the Novelty of Musical Improvisation: Evidence from an MRI Study on Composers." *Frontiers in Neuroscience* 11 (2017): 695.

Lu, J., H. Yang, X. Zhang, H. He, C. Luo, and D. Yao. "The Brain Functional State of Music Creation: An fMRI Study of Composers." *Scientific Reports* 5 (2015): 12277.

Martin, R. *Beethoven's Hair: An Extraordinary Historical Odyssey and a Scientific Mystery Solved.* New York: Crown, 2001.

Oudeyer, P. Y., and F. Kaplan. "What Is Intrinsic Motivation? A Typology of Computational Approaches." *Frontiers in Neurorobotics* 1 (2007): 6.

Ozdemir, E., A. Norton, and G. Schlaug. "Shared and Distinct Neural Correlates of Singing and Speaking." *NeuroImage* 33 (2006): 628–35.

Shen, L., A. Fishbach, and C. K. Hsee. "The Motivating-Uncertainty Effect: Uncertainty Increases Resource Investment in the Process of Reward Pursuit." *Journal of Consumer Research* 41 (2015): 1301–15.

Woolgar, A., J. Duncan, F. Manes, and E. Fedorenko. "The Multiple-Demand System but Not the Language System Supports Fluid Intelligence." *Nature Human Behaviour* 2 (2018): 200–204.

THIRD MOVEMENT

Bieszczad, K. M., and N. M. Weinberger. "Learning Strategy Trumps Motivational Level in Determining Learning-Induced Auditory Cortical Plasticity." *Neurobiology of Learning and Memory* 93 (2010): 229–39.

Duckworth, A. L., C. Peterson, M. D. Matthews, and D. R. Kelly. "Grit: Perseverance and Passion for Long-Term Goals." *Journal of Personality and Social Psychology* 92 (2007): 1087–101.

Gruber, M. J., B. D. Gelman, and C. Ranganath. "States of Curiosity Modulate Hippocampus-Dependent Learning via the Dopaminergic Circuit." *Neuron* 84 (2014): 486–96.

Kang, M. J., M. Hsu, I. M. Krajbich, G. Loewenstein, M. McClure, J. T. Wang, and C. F. Camerer. "The Wick in the Candle of Learning: Epistemic Curiosity Activates Reward Circuitry and Enhances Memory." *Psychological Science* 20 (2009): 963–73.

Moccia, L., M. Mazza, M. Di Nicola, and L. Janiri. "The Experience of Pleasure: A Perspective Between Neuroscience and Psychoanalysis." *Frontiers in Human Neuroscience* 12 (2018): 359.

Pauwels, L., S. Chalavi, and S. P. Swinnen. "Aging and Brain Plasticity." *Aging* 10 (2018): 1789–90.

Reio, T. G., Jr., and A. Wiswell. "Field Investigation of the Relationship Among Adult Curiosity, Workplace Learning, and Job Performance." *Human Resource Development Quarterly* 11 (2000): 5–30.

Rimfeld, K., Y. Kovas, P. S. Dale, and R. Plomin. "True Grit and Genetics: Predicting Academic Achievement from Personality." *Journal of Personality and Social Psychology* 111 (2016): 780–89.

Sörqvist, P., and J. E. Marsh. "How Concentration Shields Against Distraction." *Current Directions in Psychological Science* 24 (2015): 267–72.

Stahl, A. E., and L. Feigenson. "Cognitive Development: Observing the Unexpected Enhances Infants' Learning and Exploration." *Science* 348 (2015): 91–94.

Tucker-Drob, E. M., D. A. Briley, L. E. Engelhardt, F. D. Mann, and K. P. Harden. "Genetically-Mediated Associations Between Measures of Childhood Character and Academic Achievement." *Journal of Personality and Social Psychology* 111 (2016): 790–815.

Von Culin, K. R., E. Tsukayama, and A. L. Duckworth. "Unpacking Grit: Motivational Correlates of Perseverance and Passion for Long-Term Goals." *Journal of Positive Psychology* 9 (2014): 306–12.

Wade, S., and C. Kidd. "The Role of Prior Knowledge and Curiosity in Learning." *Psychonomic Bulletin and Review* 26 (2019): 1377–87.

FOURTH MOVEMENT

Ball, P. *The Music Instinct: How Music Works and Why We Can't Do Without It.* Oxford: Oxford University Press, 2010.

Bouhali, F., V. Mongelli, M. Thiebaut de Schotten, and L. Cohen. "Reading Music and Words: The Anatomical Connectivity of Musicians' Visual Cortex." *NeuroImage* 212 (2020): 116666.

Rasch, B., and J. Born. "About Sleep's Role in Memory." *Physiological Reviews* 93 (2013): 681–766.

Rudebeck, P. H., R. C. Saunders, D. A. Lundgren, and E. A. Murray. "Specialized Representations of Value in the Orbital and Ventrolateral Prefrontal Cortex: Desirability Versus Availability of Outcomes." *Neuron* 95 (2017): 1208–20.e5.

Walker, M. *Why We Sleep: Unlocking the Power of Sleep and Dreams.* New York: Scribner, 2017.

Zhang, S., P. Liu, and T. Feng. "To Do It Now or Later: The Cognitive Mechanisms and Neural Substrates Underlying Procrastination." *Wiley Interdisciplinary Reviews Cognitive Science* 10 (2019): e1492.

FIFTH MOVEMENT

Altman, J. "Are New Neurons Formed in the Brains of Adult Mammals?" *Science* 135 (1962): 1127–28.

Bengtsson, S. L., Z. Nagy, S. Skare, L. Forsman, H. Forssberg, and F. Ullén. "Extensive Piano Practicing Has Regionally Specific Effects on White Matter Development." *Nature Neuroscience* 8 (2005): 1148–50.

Craig, G. A., S. Yoo, T. Y. Du, and J. Xiao. "Plasticity in Oligodendrocyte Lineage Progression: An OPC Puzzle on Our Nerves." *European Journal of Neuroscience* 54 (2021): 5747–61.

Dansu, D. K., S. Sauma, and P. Casaccia. "Oligodendrocyte Progenitors as Environmental Biosensors." *Seminars in Cell and Developmental Biology* 116 (2021): 38–44.

Groussard, M., F. Viader, B. Landeau, B. Desgranges, F. Eustache, and H. Platel. "The Effects of Musical Practice on Structural Plasticity:

The Dynamics of Grey Matter Changes." *Brain and Cognition* 90 (2014): 174–80.

Heaney, C. F., and J. W. Kinney. "Role of GABAB Receptors in Learning and Memory and Neurological Disorders." *Neuroscience and Biobehavioral Reviews* 63 (2016): 1–28.

Hund-Georgiadis, M., and D. Y. von Cramon. "Motor-Learning-Related Changes in Piano Players and Non-musicians Revealed by Functional Magnetic-Resonance Signals." *Experimental Brain Research* 125 (1999): 417–25.

Hyde, K. L., J. Lerch, A. Norton, M. Forgeard, E. Winner, A. C. Evans, and G. Schlaug. "Musical Training Shapes Structural Brain Development." *Journal of Neuroscience* 29 (2009): 3019–25.

Kempermann, G. "Adult Neurogenesis: An Evolutionary Perspective." *Cold Spring Harbor Perspectives in Biology* 8 (2015): a018986.

Kennedy, M. B. "Synaptic Signaling in Learning and Memory." *Cold Spring Harbor Perspectives in Biology* 8 (2013): a016824.

Lee, D. J., Y. Chen, and G. Schlaug. "Corpus Callosum: Musician and Gender Effects." *NeuroReport* 14 (2003): 205–9.

Moore, E., R. S. Schaefer, M. E. Bastin, N. Roberts, and K. Overy. "Diffusion Tensor MRI Tractography Reveals Increased Fractional Anisotropy (FA) in Arcuate Fasciculus Following Music-Cued Motor Training." *Brain and Cognition* 116 (2017): 40–46.

Oechslin, M. S., C. Descloux, A. Croquelois, J. Chanal, D. Van De Ville, F. Lazeyras, and C. E. James. "Hippocampal Volume Predicts Fluid Intelligence in Musically Trained People." *Hippocampus* 23 (2013): 552–58.

Owji, S., and M. M. Shoja. "The History of Discovery of Adult Neurogenesis." *Clinical Anatomy* 33 (2020): 41–55.

Oztürk, A. H., B. Taşçioglu, M. Aktekin, Z. Kurtoglu, and I. Erden. "Morphometric Comparison of the Human Corpus Callosum in Professional Musicians and Non-musicians by Using in Vivo Magnetic Resonance Imaging." *Journal of Neuroradiology* 29 (2002): 29–34.

Penhune, V. B. "Musical Expertise and Brain Structure: The Causes and Consequences of Training." In *The Oxford Handbook of Music and the Brain*, ed. M. H. Thaut and D. A. Hodges, 419–38. Oxford: Oxford University Press, 2019.

Pytte, C. L., S. George, S. Korman, E. David, D. Bogdan, and J. R. Kirn. "Adult Neurogenesis Is Associated with the Maintenance of a Stereotyped, Learned Motor Behavior." *Journal of Neuroscience* 32 (2012): 7052–57.

Ragert, P., A. Schmidt, E. Altenmüller, and H. R. Dinse. "Superior Tactile Performance and Learning in Professional Pianists: Evidence for Meta-Plasticity in Musicians." *European Journal of Neuroscience* 19 (2004): 473–78.

Rosenkranz, K., A. Williamon, and J. C. Rothwell. "Motorcortical Excitability and Synaptic Plasticity Is Enhanced in Professional Musicians." *Journal of Neuroscience* 27 (2007): 5200–5206.

Schlaug, G., L. Jäncke, Y. Huang, J. F. Staiger, and H. Steinmetz. "Increased Corpus Callosum Size in Musicians." *Neuropsychologia* 33 (1995): 1047–55.

Swaminathan, S., and E. G. Schellenberg. "Music Training and Cognitive Abilities: Associations, Causes, and Consequences." In *The Oxford Handbook of Music and the Brain*, ed. M. H. Thaut and D. A. Hodges, 645–70. Oxford: Oxford University Press, 2019.

Vaquero, L., K. Hartmann, P. Ripollés, N. Rojo, J. Sierpowska, C. François, E. Càmara, F. T. van Vugt, B. Mohammadi, A. Samii, T. F. Münte, A. Rodríguez-Fornells, and E. Altenmüller. "Structural Neuroplasticity in Expert Pianists Depends on the Age of Musical Training Onset." *NeuroImage* 126 (2016): 106–19.

Watanabe, D., T. Savion-Lemieux, and V. B. Penhune. "The Effect of Early Musical Training on Adult Motor Performance: Evidence for a Sensitive Period in Motor Learning." *Experimental Brain Research* 176 (2007): 332–40.

Zalc, B., D. Goujet, and D. Colman. "The Origin of the Myelination Program in Vertebrates." *Current Biology* 18(2008): R511–12.

SIXTH MOVEMENT

Dunbar, R. I., K. Kaskatis, I. MacDonald, and V. Barra. "Performance of Music Elevates Pain Threshold and Positive Affect: Implications for the Evolutionary Function of Music." *Evolutionary Psychology* 10 (2012): 688–702.

Gallese, V. "The Manifold Nature of Interpersonal Relations: The Quest for a Common Mechanism." *Philosophical Transactions of the Royal Society B Biological Sciences* 358 (2003): 517–28.

Gallese, V., L. Fadiga, L. Fogassi, and G. Rizzolatti. "Action Recognition in the Premotor Cortex." *Brain* 119 (1996): 593–609.

Hou, Y., B. Song, Y. Hu, Y. Pan, and Y. Hu. "The Averaged Inter-Brain Coherence Between the Audience and a Violinist Predicts the Popularity of Violin Performance." *NeuroImage* 211 (2020): 116655.

Martín-Fernández, J., I. Burunat, C. Modroño, J. L. González-Mora, and J. Plata-Bello. "Music Style Not Only Modulates the Auditory Cortex, but Also Motor Related Areas." *Neuroscience* 457 (2021): 88–102.

Molnar-Szakacs, I., and K. Overy. "Music and Mirror Neurons: From Motion to 'E'motion." *Social Cognitive and Affective Neuroscience* 1 (2006): 235–41.

Nusseck, M., M. Zander, and C. Spahn. "Music Performance Anxiety in Young Musicians: Comparison of Playing Classical or Popular Music." *Medical Problems of Performing Artists* 30 (2015): 30–37.

Rizzolatti, G., and L. Fogassi. "The Mirror Mechanism: Recent Findings and Perspectives." *Philosophical Transactions of the Royal Society B Biological Sciences* 369 (2014): 20130420.

Rojiani, R., X. Zhang, A. Noah, and J. Hirsch. "Communication of Emotion via Drumming: Dual-Brain Imaging with Functional Near-Infrared Spectroscopy." *Social Cognitive and Affective Neuroscience* 13 (2018): 1047–57.

Stiller, A. "Toward a Biology of Music." *OPUS* 35 (1987): 12–15.

Swarbrick, D., D. Bosnyak, S. R. Livingstone, J. Bansal, S. Marsh-Rollo, M. H. Woolhouse, and L. J. Trainor. "How Live Music Moves Us: Head Movement Differences in Audiences to Live Versus Recorded Music." *Frontiers in Psychology* 9 (2019): 2682.

Weinstein, D., J. Launay, E. Pearce, R. I. Dunbar, and L. Stewart. "Group Music Performance Causes Elevated Pain Thresholds and Social Bonding in Small and Large Groups of Singers." *Evolution and Human Behavior* 37 (2016): 152–58.

Wolf, R. K. "Embodiment and Ambivalence: Emotion in South Asian Muharram Drumming." *Yearbook for Traditional Music* 32 (2000): 81–116.

Zatorre, R. J., J. L. Chen, and V. B. Penhune. "When the Brain Plays Music: Auditory-Motor Interactions in Music Perception and Production." *Nature Reviews Neuroscience* 8 (2007): 547–58.

SEVENTH MOVEMENT

Allen, E. J., P. C. Burton, C. A. Olman, and A. J. Oxenham. "Representations of Pitch and Timbre Variation in Human Auditory Cortex." *Journal of Neuroscience* 37 (2017): 1284–93.

Baumann, S., S. Koeneke, C. F. Schmidt, M. Meyer, K. Lutz, and L. Jancke. "A Network for Audio-Motor Coordination in Skilled Pianists and Non-musicians." *Brain Research* 1161 (2007): 65–78.

Bianco, R., G. Novembre, P. E. Keller, S. G. Kim, F. Scharf, A. D. Friederici, A. Villringer, and D. Sammler. "Neural Networks for Harmonic Structure in Music Perception and Action." *NeuroImage* 142 (2016): 454–64.

Blood, A. J., and R. J. Zatorre. "Intensely Pleasurable Responses to Music Correlate with Activity in Brain Regions Implicated in Reward and Emotion." *Proceedings of the National Academy of Sciences of the United States of America* 98 (2001): 11818–23.

Brattico, E., V. Alluri, B. Bogert, T. Jacobsen, N. Vartiainen, S. Nieminen, and M. Tervaniemi. "A Functional MRI Study of Happy and Sad Emotions in Music with and Without Lyrics." *Frontiers in Psychology* 2 (2011): 308.

Cross, I. "Musicality and the Human Capacity for Culture." *Musicae Scientiae* 12 (2008): 147–67.

Crowder, R. G. "Perception of the Major/Minor Distinction: I. Historical and Theoretical Foundations" *Psychomusicology: A Journal of Research in Music Cognition* 4 (1984): 3–12.

Crowder, R. G., J. S. Reznick, and S. L. Rosenkrantz. "Perception of the Major/Minor Distinction: V. Preferences Among Infants." *Bulletin of the Psychonomic Society* 29 (1991): 187–88.

Fernández-Miranda, J. C., Y. Wang, S. Pathak, L. Stefaneau, T. Verstynen, and F. C. Yeh. "Asymmetry, Connectivity, and Segmentation of the Arcuate Fascicle in the Human Brain." *Brain Structure and Function* 220 (2015): 1665–80.

Fernandez, N. B., W. J. Trost, and P. Vuilleumier. "Brain Networks Mediating the Influence of Background Music on Selective Attention." *Social Cognitive and Affective Neuroscience* 14 (2019): 1441–52.

Giovannelli, F., C. Banfi, A. Borgheresi, E. Fiori, I. Innocenti, S. Rossi, G. Zaccara, M. P. Viggiano, and M. Cincotta. "The Effect of Music on Corticospinal Excitability Is Related to the Perceived Emotion: A Transcranial Magnetic Stimulation Study." *Cortex* 49 (2013): 702–10.

Harris, R., and B. M. de Jong. "Cerebral Activations Related to Audition-Driven Performance Imagery in Professional Musicians." *PLoS One* 9 (2014): e93681.

Hasegawa, T., K. Matsuki, T. Ueno, Y. Maeda, Y. Matsue, Y. Konishi, and N. Sadato. "Learned Audio-Visual Cross-Modal Associations in Observed Piano Playing Activate the Left Planum Temporale. An fMRI Study." *Cognitive Brain Research* 20 (2004): 510–18.

Hunter, P. G., and E. G. Schellenberg. "Music and Emotion." Music Perception 36 (2010): 129–64.

Jäncke, L. "Music, Memory and Emotion." *Journal of Biology* 7 (2008): 21.

Janzen, T. B., and M. H. Thaut. "Cerebral Organization of Music Processing." In *The Oxford Handbook of Music and the Brain*, ed. M. H. Thaut and D. A. Hodges, 89–121. Oxford: Oxford University Press, 2019.

Kastner, M. P., and R. G. Crowder. "Perception of the Major/Minor Distinction: IV. Emotional Connotations in Young Children." *Music Perception* 8 (1990): 189–201.

Klein, M. E., and R. J. Zatorre. "A Role for the Right Superior Temporal Sulcus in Categorical Perception of Musical Chords." *Neuropsychologia* 49 (2011): 878–87.

Koelsch, S. "Brain Correlates of Music-Evoked Emotions." *Nature Reviews Neuroscience* 15 (2014): 170–80.

Koelsch, S. "Toward a Neural Basis of Music Perception—A Review and Updated Model." *Frontiers in Psychology* 2 (2011): 110.

Koelsch, S., T. Fritz, D. Y. v. Cramon, K. Müller, and A. D. Friederici. "Investigating Emotion with Music: An fMRI Study." *Human Brain Mapping* 27 (2006): 239–50.

Koelsch, S., and S. Skouras. "Functional Centrality of Amygdala, Striatum and Hypothalamus in a 'Small-World' Network Underlying Joy: An fMRI Study with Music." *Human Brain Mapping* 35 (2014): 3485–98.

Koelsch, S., S. Skouras, T. Fritz, P. Herrera, C. Bonhage, M. B. Küssner, and A. M. Jacobs. "The Roles of Superficial Amygdala and Auditory Cortex in Music-Evoked Fear and Joy." *NeuroImage* 81 (2013): 49–60.

Kolchinsky, A., N. Dhande, K. Park, and Y. Y. Ahn. "The Minor Fall, the Major Lift: Inferring Emotional Valence of Musical Chords Through Lyrics." *Royal Society Open Science* 4 (2017): 170952.

Lehne, M., M. Rohrmeier, and S. Koelsch. "Tension-Related Activity in the Orbitofrontal Cortex and Amygdala: An fMRI Study with Music." *Social Cognitive and Affective Neuroscience* 9 (2014): 1515–23.

Michaelis, K., M. Wiener, and J. C. Thompson. "Passive Listening to Preferred Motor Tempo Modulates Corticospinal Excitability." *Frontiers in Human Neuroscience* 8 (2014): 252.

Mitterschiffthaler, M. T., C. H. Fu, J. A. Dalton, C. M. Andrew, and S. C. Williams. "A Functional MRI Study of Happy and Sad Affective States Induced by Classical Music." *Human Brain Mapping* 28 (2007): 1150–62.

Molinari, M., M. G. Leggio, V. Filippini, M. C. Gioia, A. Cerasa, and M. H. Thaut. "Sensorimotor Transduction of Time Information Is Preserved in Subjects with Cerebellar Damage." *Brain Research Bulletin* 67 (2005): 448–58.

Moreno, S., Y. Lee, M. Janus, and E. Bialystok. "Short-Term Second Language and Music Training Induces Lasting Functional Brain Changes in Early Childhood." *Child Development* 86 (2015): 394–406.

Nozaradan, S., M. Schwartze, C. Obermeier, and S. A. Kotz. "Specific Contributions of Basal Ganglia and Cerebellum to the Neural Tracking of Rhythm." *Cortex* 95 (2017): 156–68.

Pallesen, K. J., E. Brattico, C. Bailey, A. Korvenoja, J. Koivisto, A. Gjedde, and S. Carlson. "Emotion Processing of Major, Minor, and Dissonant Chords: A Functional Magnetic Resonance Imaging Study." *Annals of the New York Academy of Sciences* 1060 (2005): 450–53.

Paquette, S., S. Fujii, H. C. Li, and G. Schlaug. "The Cerebellum's Contribution to Beat Interval Discrimination." *NeuroImage* 163 (2017): 177–82.

Phillips, T., and A. D'Angour. *Music, Text, and Culture in Ancient Greece.* Oxford: Oxford University Press, 2018.

Pietschnig, J., M. Voracek, and A. K. Formann. "Mozart Effect—Shmozart Effect: A Meta-Analysis." *Intelligence* 38 (2010): 314–23.

Putkinen, V., S. Nazari-Farsani, K. Seppälä, T. Karjalainen, L. Sun, H. K. Karlsson, M. Hudson, T. T. Heikkilä, J. Hirvonen, and L. Nummenmaa. "Decoding Music-Evoked Emotions in the Auditory and Motor Cortex." *Cerebral Cortex* 31 (2021): 2549–60.

Rauscher, F. H., G. L. Shaw, and K. N. Ky. "Music and Spatial Task Performance." *Nature* 365 (1993): 611.

Salimpoor, V. N., M. Benovoy, K. Larcher, A. Dagher, and R. J. Zatorre. "Anatomically Distinct Dopamine Release During Anticipation and Experience of Peak Emotion to Music." *Nature Neuroscience* 14 (2011): 257–62.

Salimpoor, V. N., I. van den Bosch, N. Kovacevic, A. R. McIntosh, A. Dagher, and R. J. Zatorre. "Interactions Between the Nucleus Accumbens and Auditory Cortices Predict Music Reward Value." *Science* 340 (2013): 216–19.

Samson, S. "Neuropsychological Studies of Musical Timbre." *Annals of the New York Academy of Sciences* 999 (2003): 144–51.

Stupacher, J., M. J. Hove, G. Novembre, S. Schütz-Bosbach, and P. E. Keller. "Musical Groove Modulates Motor Cortex Excitability: A TMS Investigation." *Brain and Cognition* 82 (2013): 127–36.

Suzuki, M., N. Okamura, Y. Kawachi, M. Tashiro, H. Arao, T. Hoshishiba, J. Gyoba, and K. Yanai. "Discrete Cortical Regions Associated with the Musical Beauty of Major and Minor Chords." *Cognitive, Affective, and Behavioral Neuroscience* 8 (2008): 126–31.

Thompson, W. F., E. G. Schellenberg, and G. Husain. "Arousal, Mood, and the Mozart Effect." *Psychological Science* 12 (2001): 248–51.

Trainor, L. J. "Are There Critical Periods for Musical Development?" *Developmental Psychobiology* 46 (2005): 262–78.

Trehub, S. E. "Musical Predispositions in Infancy." *Annals of the New York Academy of Sciences* 930 (2001): 1–16.

Wallin, N. L., B. Merker, and S. Brown, eds. *The Origins of Music.* Cambridge, MA: MIT Press, 2000.

Wessinger, C. M., M. H. Buonocore, C. L. Kussmaul, and G. R. Mangun. "Tonotopy in Human Auditory Cortex Examined with Functional Magnetic Resonance Imaging." *Human Brain Mapping* 5 (1997): 18–25.

Wuttke-Linnemann, A., U. M. Nater, U. Ehlert, and B. Ditzen. "Sex-Specific Effects of Music Listening on Couples' Stress in Everyday Life." *Scientific Reports* 9 (2019): 4880.

Zatorre, R. J., and V. N. Salimpoor. "From Perception to Pleasure: Music and Its Neural Substrates." *Proceedings of the National Academy of Sciences of the United States of America* 110 (2013): 10430–37.

EIGHTH MOVEMENT

Ahrends, C. "Does Excessive Music Practicing Have Addiction Potential?" *Psychomusicology: Music, Mind, and Brain* 27 (2017): 191–202.

Belfi, A. M., and P. Loui. "Musical Anhedonia and Rewards of Music Listening: Current Advances and a Proposed Model." *Annals of the New York Academy of Sciences* 1464 (2020): 99–114.

Berridge, K. C., and M. L. Kringelbach. "Pleasure Systems in the Brain." *Neuron* 86 (2015): 646–64.

Berridge, K. C., and T. E. Robinson. "Liking, Wanting, and the Incentive-Sensitization Theory of Addiction." *American Psychology* 71 (2016): 670–79.

Boer, D., R. Fischer, M. Strack, M. H. Bond, E. Lo, and J. Lam. "How Shared Preferences in Music Create Bonds Between People: Values as the Missing Link." *Personality and Social Psychology Bulletin* 37 (2011): 1159–71.

Bonneville-Roussy, A., P. J. Rentfrow, M. K. Xu, and J. Potter. "Music Through the Ages: Trends in Musical Engagement and Preferences from Adolescence Through Middle Adulthood." *Journal of Personality and Social Psychology* 105 (2013): 703–17.

DiFeliceantonio, A. G., O. S. Mabrouk, R. T. Kennedy, and K. C. Berridge. "Enkephalin Surges in Dorsal Neostriatum as a Signal to Eat." *Current Biology* 22 (2012): 1918–24.

Dubé, L., and J. Le Bel. "The Content and Structure of Laypeople's Concept of Pleasure." *Cognition and Emotion* 17 (2003): 263–95.

Eerola, T., J. K. Vuoskoski, and H. Kautiainen. "Being Moved by Unfamiliar Sad Music Is Associated with High Empathy." *Frontiers in Psychology* 7 (2016): 1176.

Egermann, H., and S. McAdams. "Empathy and Emotional Contagion as a Link Between Recognized and Felt Emotions in Music Listening." *Music Perception: An Interdisciplinary Journal* 31 (2013): 139–56.

Ferreri, L., E. Mas-Herrero, R. J. Zatorre, P. Ripollés, A. Gomez-Andres, H. Alicart, G. Olivé, J. Marco-Pallarés, R. M. Antonijoan, M. Valle, J. Riba, and A. Rodriguez-Fornells. "Dopamine Modulates the Reward Experiences Elicited by Music." *Proceedings of the National Academy of Sciences of the United States of America* 116 (2019): 3793–98.

Garrido, S., and E. Schubert. "Individual Differences in the Enjoyment of Negative Emotion in Music: A Literature Review and Experiment." *Music Perception* 28 (2011): 279–96.

Green, B., D. Kavanagh, and R. Young. "Being Stoned: A Review of Self-Reported Cannabis Effects." *Drug and Alcohol Review* 22 (2003): 453–60.

Greenberg, D. M., S. Baron-Cohen, D. J. Stillwell, M. Kosinski, and P. J. Rentfrow. "Musical Preferences Are Linked to Cognitive Styles." *PLoS One* 10 (2015): e0131151.

Harbaugh, W. T., U. Mayr, and D. R. Burghart. "Neural Responses to Taxation and Voluntary Giving Reveal Motives for Charitable Donations." *Science* 316 (2007): 1622–25.

Harris, R. J., S. J. Hoekstra, C. L. Scott, F. W. Sanborn, J. A. Karafa, and J. D. Brandenburg. "Young Men's and Women's Different Autobiographical Memories of the Experience of Seeing Frightening Movies on a Date." *Media Psychology* 2 (2000): 245–68.

Huron, D., and J. K. Vuoskoski. "On the Enjoyment of Sad Music: Pleasurable Compassion Theory and the Role of Trait Empathy." *Frontiers in Psychology* 11 (2020): 1060.

Juslin, P. N., and P. Laukka. "Expression, Perception, and Induction of Musical Emotions: A Review and a Questionnaire Study of Everyday Listening." *Journal of New Music Research* 33 (2004): 217–38.

Kaneshiro, B., F. Ruan, C. W. Baker, and J. Berger. "Characterizing Listener Engagement with Popular Songs Using Large-Scale Music Discovery Data." *Frontiers in Psychology* 8 (2017): 416.

Levinson, J. *Music, Art and Metaphysics: Essays in Philosophical Aesthetics*. Oxford: Oxford University Press, 1990.

Lim, M. S., M. E. Hellard, J. S. Hocking, and C. K. Aitken. "A Cross-Sectional Survey of Young People Attending a Music Festival: Associations Between Drug Use and Musical Preference." *Drug and Alcohol Review* 27 (2008): 439–41.

Loui, P., S. Patterson, M. E. Sachs, Y. Leung, T. Zeng, and E. Przysinda. "White Matter Correlates of Musical Anhedonia: Implications for Evolution of Music." *Frontiers in Psychology* 8 (2017): 1664.

Mallik, A., M. L. Chanda, and D. J. Levitin. "Anhedonia to Music and Mu-Opioids: Evidence from the Administration of Naltrexone." *Scientific Reports* 7 (2017): 41952.

Martínez-Molina, N., E. Mas-Herrero, A. Rodríguez-Fornells, R. J. Zatorre, and J. Marco-Pallarés. "Neural Correlates of Specific Musical Anhedonia." *Proceedings of the National Academy of Sciences of the United States of America* 113 (2016): E7337–E45.

Mas-Herrero, E., M. Karhulahti, J. Marco-Pallares, R. J. Zatorre, and A. Rodriguez-Fornells. "The Impact of Visual Art and Emotional Sounds in Specific Musical Anhedonia." *Progress in Brain Research* 237 (2018): 399–413.

McDermott, J. H., A. F. Schultz, E. A. Undurraga, and R. A. Godoy. "Indifference to Dissonance in Native Amazonians Reveals Cultural Variation in Music Perception." *Nature* 535 (2016): 547–50.

McPherson, M. J., S. E. Dolan, A. Durango, T. Ossandon, J. Valdés, E. A. Undurraga, N. Jacoby, R. A. Godoy, and J. H. McDermott. "Perceptual Fusion of Musical Notes by Native Amazonians Suggests Universal Representations of Musical Intervals." *Nature Communications* 11 (2020): 2786.

Miu, A. C., and F. R. Balteş. "Empathy Manipulation Impacts Music-Induced Emotions: A Psychophysiological Study on Opera." *PloS One* 7 (2012): e30618.

North, A. C. "Individual Differences in Musical Taste." *American Journal of Psychology* 123 (2010): 199–208.

Omigie, D., and J. Ricci. "Accounting for Expressions of Curiosity and Enjoyment During Music Listening." *Psychology of Aesthetics, Creativity, and the Arts*, advance online publication, 2022.

Palamar, J. J., M. Griffin-Tomas, and D. C. Ompad. "Illicit Drug Use Among Rave Attendees in a Nationally Representative Sample of U.S. High School Seniors." *Drug and Alcohol Dependence* 152 (2015): 24–31.

Rabinowitch, T. C., I. Cross, and P. Burnard. "Long-Term Musical Group Interaction Has a Positive Influence on Empathy in Children." *Psychology of Music* 41 (2013): 484–98.

Rentfrow, P. J., and S. D. Gosling. "The Do Re Mi's of Everyday Life: The Structure and Personality Correlates of Music Preferences." *Journal of Personality and Social Psychology* 84 (2003): 1236–56.

Rentfrow, P. J., and J. A. McDonald. "Music Preferences and Personality." In *Handbook of Music and Emotion*, ed. P. N. Juslin and J. Sloboda, 669–95. Oxford: Oxford University Press, 2010.

Reybrouck, M., P. Podlipniak, and D. Welch. "Music Listening as Coping Behavior: From Reactive Response to Sense-Making." *Behavioral Science* 10 (2020): 119.

Sachs, M. E., A. Damasio, and A. Habibi. "The Pleasures of Sad Music: A Systematic Review." *Frontiers in Human Neuroscience* 9 (2015): 404.

Salimpoor, V. N., M. Benovoy, K. Larcher, A. Dagher, and R. J. Zatorre. "Anatomically Distinct Dopamine Release During Anticipation and Experience of Peak Emotion to Music." *Nature Neuroscience* 14 (2011): 257–62.

Schäfer, T., P. Sedlmeier, C. Städtler, and D. Huron. "The Psychological Functions of Music Listening." *Frontiers in Psychology* 4 (2013): 511.

Schmuziger, N., J. Patscheke, R. Stieglitz, and R. Probst. "Is There Addiction to Loud Music? Findings in a Group of Non-professional Pop/Rock Musicians." *Audiology Research* 2 (2012): e11.

Sihvonen, A. J., T. Särkämö, A. Rodríguez-Fornells, P. Ripollés, T. F. Münte, and S. Soinila. "Neural Architectures of Music—Insights from Acquired Amusia." *Neuroscience and Biobehavioral Reviews* 107 (2019): 104–14.

Stone, N. L., S. A. Millar, P. J. J. Herrod, D. A. Barrett, C. A. Ortori, V. A. Mellon, and S. E. O'Sullivan. "An Analysis of Endocannabinoid Concentrations and Mood Following Singing and Exercise in Healthy Volunteers." *Frontiers in Behavioral Neuroscience* 12 (2018): 269.

Tart, C. T. "Marijuana Intoxication Common Experiences." *Nature* 226 (1970): 701–4.

Taruffi, L., and S. Koelsch. "The Paradox of Music-Evoked Sadness: An Online Survey." *PLoS One* 9 (2014): e110490.

Van den Tol, A. J. M., and J. Edwards. "Exploring a Rationale for Choosing to Listen to Sad Music When Feeling Sad." *Psychology of Music* 41 (2013): 440–65.

Van Havere, T., W. Vanderplasschen, J. Lammertyn, E. Broekaert, and M. Bellis. "Drug Use and Nightlife: More Than Just Dance Music." *Substance Abuse Treatment Prevention and Policy* 6 (2011): 18.

Vuoskoski, J. K., and T. Eerola. "Can Sad Music Really Make You Sad? Indirect Measures of Affective States Induced by Music and Autobiographical Memories." *Psychology of Aesthetics, Creativity, and the Arts* 6 (2012): 204.

Vuoskoski, J. K., and T. Eerola. "The Role of Mood and Personality in the Perception of Emotions Represented by Music." *Cortex* 47 (2011): 1099–106.

Vuoskoski, J. K., W. F. Thompson, D. McIlwain, and T. Eerola. "Who Enjoys Listening to Sad Music and Why?" *Music Perception* 29 (2012): 311–17.

Zald, D., and R. J. Zatorre. "Music." In *Neurobiology of Sensation and Reward*, ed. D. Zald, R. J. Zatorre, and J. A. Gottfried. Boca Raton, FL: CRC Press/Taylor and Francis, 2011.

CODA

Meilandt, W. J., E. Barea-Rodriguez, S. A. Harvey, and J. L. Martinez Jr. "Role of Hippocampal CA3 Mu-Opioid Receptors in Spatial Learning and Memory." *Journal of Neuroscience* 24 (2004): 2953–62.

INDEX

Note: Page numbers followed by *f* indicate a figure on the corresponding page.

INDEX

Milton Keynes UK
Ingram Content Group UK Ltd.
UKHW011429140124
435941UK00005B/7/J

9 780231 209106